THE POLITICS OF ENVIRONMENTAL DISCOURSE

The Politics of Environmental Discourse

Ecological Modernization and the Policy Process

MAARTEN A. HAJER

CLARENDON PRESS · OXFORD

1995

Oxford University Press, Walton Street, Oxford OX2 6DP
Oxford   New York
Athens Auckland Bangkok Bombay
Calcutta Cape Town Dar es Salaam Delhi
Florence Hong Kong Istanbul Karachi
Kuala Lumpur Madras Madrid Melbourne
Mexico City Nairobi Paris Singapore
Taipei Tokyo Toronto
and associated companies in
Berlin Ibadan

Oxford is a trade mark of Oxford University Press

Published in the United States
by Oxford University Press Inc., New York

British Library Cataloguing in Publication Data
Data available

Library of Congress Cataloging in Publication Data
Hajer, Maarten A., 1962–
The politics of environmental discourse : ecological modernization
and the policy process / Maarten A. Hajer.
Includes bibliographical references.
1. Environmental policy. 2. Ecology. 3. Economic development–
Environmental aspects. I. Title
GE170.H36 1995 363.7–dc20 95–5877
ISBN 0–19–827969–8

1 3 5 7 9 10 8 6 4 2

Typeset by Graphicraft Typesetters Ltd., Hong Kong
Printed in Great Britain
on acid-free paper by
Biddles Ltd., Guildford and King's Lynn

ACKNOWLEDGEMENTS

This book originates in my graduate research at the University of Oxford. I would therefore first of all like to express my gratitude to all those that have made my stay there possible. For financial support I am indebted to the British Foreign and Commonwealth Office for granting me a Chevening Scholarship during those years. I would also like to mention the British Council that administered the scholarship scheme and have always kept an active interest in my personal and intellectual well-being. Many thanks are due to the Institute for Law and Public Policy at the University of Leyden where I held a post as a research fellow at a later stage of my research. I would like to thank the consecutive directors of the Institute, Wim Derksen and Nick Huls, for their interest and continued support. For their secretarial support I would like to thank Anne-Marie Krens and Thea de Beer of the Leiden Institute.

In the course of my research I have begged for favours from many people. To start with I would like to thank the many people that agreed to be interviewed although time was often extremely precious for them. I would also like to express my gratitude to those librarians and archivists who helped me to gather the empirical material for this book.

Special thanks are due to all those who have contributed to my thinking during the writing of the book. In particular I would like to extend my thanks to David Harvey, my thesis supervisor at Oxford, who over the years greatly influenced my thinking, probably more than I am now aware of. I would also like to thank Willem Trommel, Hein-Anton van der Heijden, Frank Fischer, Patsy Healey, Roland Bal, Tim Marshall, Iain McLean, and Thijs Drupsteen, who all read and criticized substantial sections of earlier versions of this book. I owe special thanks to Michiel Schwarz and Anton Hemerijck, who both read the entire manuscript and were a constant source of stimulating suggestions and criticisms. Our discussions have helped me tremendously in clarifying the ideas presented in this book.

While preparing the final manuscript of this book I benefited enormously from the written comments of Albert Weale, and from the report of the anonymous reviewer for OUP, as well as from various discussions with my collegues at the Institute of Sociology at the University of Munich, Germany, in particular Ulrich Beck, Ronald Hitzler, and Elmar Koenen. Thanks are due to all these people.

Finally, I would like to express my gratitude to Janine and Minke for their endurance and support.

M.A.H.

Munich/Amsterdam
Summer 1994

CONTENTS

LIST OF FIGURE AND TABLES

ABBREVIATIONS

AMvB	Algemene Maatregel van Bestuur (government directive)
ANP	Algemeen Nederlands Persbureau
AVO	Additioneel Verzuringsonderzoek (Dutch Priority Programme on Acidification)
BATNEEC	Best Available Technology Not Entailing Excessive Cost
BPM	Best Practicable Means
CBI	Confederation of British Industry
CDA	Christen-democratisch Appel (Dutch christian-democratic party)
CEC	Commission of the European Communities
CEGB	Central Electricity Generating Board
CPPU	Central Policy Planning Unit (DoE)
CRMH	Centrale Raad voor de Milieuhygiëne
DAS	Dutch Acidification Simulation (computer model)
DoE	Department of the Environment
D'66	Democraten 1966 (Dutch progressive liberal party)
EC	European Community
EIA	Environmental Impact Assessment
EMEP	European Monitoring and Evaluation Programme (UN–ECE)
EPR	Environmental Performance Review
FAO	Food and Agriculture Organization (UN)
FC	Forestry Commission
FGD	Flue Gas Desulphurization
FoE	Friends of the Earth
IIASA	International Institute for Applied Systems Analysis
IMP	Indicatief Meerjaren Programma (Dutch policy plan)
IPCC	Intergovernmental Panel on Climate Change
Ir.	Official title for Dutch engineers
ITE	Institute for Terrestrial Ecology
IUCN	International Union for the Conservation of Nature and Natural Resources

IVEM	Interfacultaire Vakgroep Energie en Milieukunde (University of Groningen)
LRTAP	Long-Range Transboundary Air Pollution Convention of 1979 ('Geneva Convention')
MIT	Massachusetts Institute of Technology
MoE	Minutes of Evidence (Parliamentary Select Committee)
NCC	Nature Conservancy Council
NERA	National Economic Research Associates
NERC	National Environment Research Council
NGO	Non-Governmental Organization
NH_3	Ammonia
NMP	Nationaal Milieubeleids Plan (Dutch environmental policy plan)
NOS	Nederlandse Omroep Stichting (Dutch Broadcasting Association)
NO_x	Nitrogen Oxides
OECD	Organization for Economic Co-operation and Development
PSP	Pacifistisch Socialistische Partij (Dutch radical progressive party)
PvdA	Partij van de Arbeid (Dutch labour party)
RAINS	Regional Acidification Information and Simulation (IIASA model)
RIN	Rijks Instituut voor Natuurbeheer (Dutch National Research Institute for Conservation)
RIVM	Rijks Instituut voor Volksgezondheid en Milieuhygiëne (Dutch National Institute for Public Health and Environmental Protection)
RSPB	Royal Society for the Protection of Birds
SEP	Samenwerkende Electriciteits Producenten (Dutch Federation of Electricity producers)
SER	Sociaal Economische Raad (Social-economic Council, NL)
SPSS	Statistical Package for the Social Sciences
SO_2	Sulphur dioxide
SPA	Science-based Policy Approach (UK)
SWAP	Surface Waters Acidification Programme
UCV	Uitgebreide Commissie Vergadering (Parliamentary Select Committee meeting, NL)

UK-CEED	United Kingdom Centre for Economic and Environmental Development
UN-ECE	United Nations Economic Commission for Europe
UNEP	United Nations Environmental Programme
VNO	Vereniging van Nederlandse Ondernemingen (Confederation of Dutch Industries)
VOCs	Volatile Organic Compounds
VoMil	Departement voor Volksgezondheid en Milieuhygiëne (Department for Public Health and Environmental Hygiene, NL)
VRL	Voorlopige Raad inzake de Luchtverontreiniging (NL)
VROM	Departement van Volkshuisvesting, Ruimtelijke Ordening en Milieuhygiëne (Department for public housing, spatial planning, and environmental hygiene, NL)
VVD	Volkspartij voor Vrijheid en Democratie (Dutch conservative party)
WABM	Wet algemene bepalingen milieuhygiëne (General Environmental Regulation Act, NL)
WCED	World Commission on Environment and Development (UN)
WCS	World Conservation Strategy (IUCN)
WHO	World Health Organization
WISE	World Information Services on Energy
WZB	Wissenschaftszentrum Berlin für Sozialforschung (Science Centre, Berlin)

Note

Handelingen Tweede Kamer: The official parliamentary papers of the Dutch government are organized by parliamentary year, which runs from September to June. Government papers not published in this form are listed by Department.

Introduction

Over the past few years 'the ecological crisis' has come to occupy a permanent place on the public agenda. It has become an issue whose importance goes without saying. One only has to refer to 'the ecological crisis' and everybody nods meaningfully as if one knows what is being referred to. The ecological crisis, so the consensus seems to be, has to be faced. Yet somehow cracks are beginning to emerge in the picture of a new ecological consensus. Here the UN Earth Summit held in Rio de Janeiro in June 1992 played a role of great symbolic importance. The global conference was meant to be the culmination of the integrative effort and was to mark the start of a new ecological era for which Agenda 21 was to point the way. As many commentators have pointed out since then, if anything Rio in actual fact showed the many conceptual holes and political ambivalences that had slumbered beneath the consensus over the need to arrive at some sort of sustainable development. Over time these differences have only become more obvious. This comes out in the struggle over the meaning of the notion of the term 'sustainable development', which functioned as the linchpin in the creation of the new consensus. As Brooks (1992: 408) has pointed out, since the Brundtland Report (re-)introduced the concept in 1987, at least forty working definitions of sustainable development have appeared. Consequently many different projects are furthered under the flag of sustainable development and quarrels have started to emerge about what sustainable development really is.

Most analyses embark on an explanation of the political interests that stand in the way of a real 'ecological turn'. This book aims to enhance our insight into environmental politics by raising a different sort of question. It argues that, if examined closely, environmental discourse is fragmented and contradictory. Environmental discourse is an astonishing collection of claims and concerns brought together by a great variety of actors. Yet somehow we distil seemingly coherent problems out of this jamboree of claims and concerns. How does this work? How do problems get defined and what sort of political consequences does this have?

To analyse this process we draw on the methodology of discourse analysis. This does not simply refer to the analysis of the discussion of the ecological crisis but examines all those factors that influence the way in which we conceive of the environmental problematique. It will be argued that the developments in environmental politics critically depend on the specific social construction of environmental problems. We do not simply analyse what is being said, but also include the institutional context in which this is done and which co-determines what can be said meaningfully.

Special focus is on the domain of environmental policy-making. The analysis of policy-making usually has a rather instrumental orientation, whereby the goal of improving the quality of the policy output guides the analytical effort. In actual fact policy-making deserves our attention as a social phenomenon in its own right. Policy-making is not just a matter of finding acceptable solutions for preconceived problems. It is also the dominant way in which modern societies regulate latent social conflicts. Yet in this policy-making involves much more than merely dreaming up clever ways of creating solutions. It requires first of all the redefinition of a given social phenomenon in such a way that one can also find solutions for them. Experts of all sorts are called upon to define the problem and its parameters. Within what domain do we have to find our solutions? What institutional commitments have to be respected? Which social conditions are malleable, which ones are fixed?

This book investigates environmental policy-making as the socially accepted set of practices through which we try to face what has become known as the ecological crisis. In so doing we seek to shed new light on the dynamics of environmental conflict in modern societies. Here much conceptual work needs to be done. The environmental conflict simply has too many component parts to be understandable if one delimits oneself to one of the many established academic domains. For instance, in order to understand the regulation of the conflict over acid rain in the 1980s, one has to open up for insights and ideas from an array of disciplines including the sociology of science, sociology of risk, policy analysis, human geography, anthropology, and political science.

Throughout the book we try to come to grips with the way in which something as big and potentially explosive as the ecological crisis is being managed through environmental policy-making. Here the ecological crisis is seen as a major challenge to the institutions of modern society. As the German sociologist Ulrich Beck reminds us, incidents like Chernobyl or Bhopal are not merely physically explosive, but are potentially socially explosive as well. As we will discuss in Chapter 1, they raise questions over the very functioning of some of the central institutions of modern societies. We should therefore seek to keep an eye on the way in which the practices of policy-making not only facilitate the regulation of a physical problem but can be seen to fulfil a much broader role in regulating the latent social conflict that is inherent in ecological matters.

The historical argument, in brief, is that a new way of conceiving environmental problems has emerged since the late 1970s. This policy discourse of ecological modernization recognizes the ecological crisis as evidence of a fundamental omission in the workings of the institutions of modern society. Yet, unlike the radical environmental movements of the 1970s, it suggests that environmental problems can be solved in accordance with the workings of the main institutional arrangements of society. Environmental management is seen as a positive-sum game: pollution prevention pays.

It is easy to see why ecological modernization would quickly conquer the hearts of politicians and policy-makers. Yet one may, of course, question some of its assumptions. On a theoretical level one can show the kind of issues and uncertainties that ecological modernization leaves unaddressed. Yet it is only through empirical work that we can come to an assessment of the effects of ecological modernization on the regulation of the environmental conflict. For that purpose this study analyses the influence of ecological modernization on the regulation of the problem of acid rain in Great Britain and the Netherlands. It reconstructs so-called discourse-coalitions that could be seen shaping up around this problem. In so doing this book seeks to show how social constructivism and discourse analysis add essential insights to our analysis of contemporary environmental politics. These insights are of great importance to the question of how society should seek to come to terms with today's ecological dilemma.

AN OUTLINE OF THE BOOK

The thesis of this book is that whether or not environmental problems appear as anomalies to the existing institutional arrangements depends first of all on the way in which these problems are framed and defined. That is what the environmental conflict is about. In this context the emergence of ecological modernization as the new dominant way of conceptualizing environmental problems becomes an important topic of analysis. It raises a long list of compelling questions: What can be thought within its structures? Where does it hit its conceptual limits? In what sense does it open up solidified relations of power? Is ecological modernization to be understood as the materialization of the original ideas of the environmental debate in its early stages, or does it signify the collapse of critical discourse? To what extent does it facilitate a more reflexive form of modernization?

Such questions cannot be answered in the abstract. To be able to come to grips with the effects of this conceptual innovation one will need to examine how the emergence and acceptance of such a conceptual language was taken up in actual practice and what sort of institutional innovations it brought about. For this reason this book contains a larger empirical section which presents a discourse analysis of the emergence of ecological modernization in the United Kingdom and the Netherlands.

Discourse analysis has been employed in a great many ways, but the operationalization of the work of authors like Michel Foucault or Rom Harré for purpose of the analysis of contemporary policy-making is rather rare. Chapter 2 seeks to use it in this way. It draws on the distinct discourse-theoretical perspectives developed by Michel Foucault and by a group of social-psychologists such as Harré and Billig. Out of the confrontation of these perspectives we derive a discourse-analytical framework for the study of contemporary public policy. To show the conceptual innovations for the study of policy-making this 'argumentative' approach is then compared with the received work on 'policy-oriented learning' of Paul Sabatier.

Chapter 3 presents an analysis of the historical circumstances out of which ecological modernization emerged. This chapter seeks to present an institutional analysis out of which the dominant position of ecological modernization might be explained. So, rather than giving yet another review of the various ideological positions that can be found in contemporary environmental politics since 1972, it seeks to present a first institutional genealogy that shows where the eco-modernist concepts were first developed and explains how they were able to gain so much in influence in the mid-1980s.

The second part of the book contains case-studies of the United Kingdom and the Netherlands. They serve to give insights into the way in which ecological modernization affected environmental policy-making. For this purpose it examines the regulation of the acid rain controversies in the two countries. They should illuminate the new practices where environmental politics is being made. The case-studies do not focus on the parliamentary debates (although they are obviously part of the analysis) but identify various other fields of practice in the environmental domain where 'sub-political processes' develop.

Acid rain was not picked at random. There are essentially two reasons for a detailed investigation. First, acid rain is a fine example of what are defined in Chapter 1 as the emblems of environmental discourse. This refers to issues that stand out in a particular period and dominate public and political attention. These issues can be shown to function as a 'metaphor' for the environmental problematique at large. That is to say, people understand the bigger problem of the ecological crisis through the example of certain emblems.

Second, acid rain has played a key role in the emergence of ecological modernization over the last twenty years, both in Europe and North America.[1] As is well known, acid rain was the issue that led the Swedish government to call for an international conference on the environment which eventually became the 1972 UN Stockholm Conference. The emission of sulphur dioxide (SO_2), nitrogen oxides (NO_x), and ammonia (NH_3) was alleged to cause the death of lakes and trees and was also held responsible for the corrosion of buildings and sculptures. After a period of relative silence in the 1970s, acid rain became widely accepted as being one of the major environmental hazards of Europe and North

---

[1] Cf. Schmandt *et al.* 1988; Regens and Rycroft 1988 for early studies on the American controversies; Boehmer-Christiansen and Skea 1991 for a comparative study on the European controversies; and Cowling 1982 and Wetstone 1987 for a treatment of the history of the issue.

America during the 1980s.[2] It was defined as an example of the new generation of environmental hazards that started to dominate the environmental agenda in the 1980s. Like the greenhouse effect and the diminishing ozone layer, acid rain is an international form of pollution, by nature more or less invisible yet cumulative in its effects. As an example of the new generation of environmental issues, acid rain served as one of the prime anomalies to the then prevailing *ad hoc* approach to environmental regulation. It was one of the central issues in the context of which the concepts of ecological modernization, like critical loads, multiple stress, or internalization of pollution prevention were first introduced and developed.

Acid rain is a nice case since it has always been beset with uncertainties. Many environmental issues receive political attention following a pollution incident (like urban air pollution after the notorious London fog of 1952). Yet as a creeping and cumulative form of pollution acid rain was a typical example of the new generation of environmental issues that, more than their predecessors, depend on their discursive creation. Furthermore acid rain is a problem for which there are many different solutions that all depend on a specific creation of the problem. Hence it is an issue that illustrates the crucial importance of recognizing multiple problems and multiple definitions in a debate.

However, if the new environmental conflict really is socially constructed we should also allow for international variation. This book is therefore conceptualized as a comparative case-study on the regulation of the problem of acid rain in two OECD countries, the UK and the Netherlands, between 1972 and 1990. Both the UK and the Netherlands were, right from the start, drawn into the heart of the European debate. Yet Britain was always seen as the prime culprit in relation to the acidification of the environment in southern Scandinavia, while the Netherlands was seen internationally as one of the countries that was most committed to helping to find a solution to the international acid-rain problem. Interestingly, the received knowledge of the policy styles of the two countries suggests widely different traditions, which makes the comparative study interesting from the point of view of the

---

[2] WCED 1987: 178. Cf. also Wetstone 1987.

potential unifying influence of the emergence of an international policy discourse like ecological modernization.

So far the Anglo-Saxon literature on the Dutch policy planning approach that was developed during the 1980s is still rather limited. The presentation of the Dutch model is nearly always based on the description of the elaborate Dutch Policy Plans but is rarely supported by an analysis of the actual practical achievements of this approach. The comparative study of the UK and the Netherlands hopes to contribute to a more sophisticated understanding of Dutch environmental politics.

Finally, Chapter 6 discusses the contribution of the social constructivist analysis of environmental policy-making and politics. It argues that the strength of social constructivism and discourse analysis is not limited to the demystification of what previously appeared as rational. Social constructivism should not only be concerned with the opening of 'black boxes' but should also help to see how new 'reflexive' institutional arrangements may be developed, and should assess to what extent one can expect such institutional arrangements to overcome the fundamental contradictions that the ecological dilemma conceals. The chapter posits global environmental problems as the emblems of the 1990s, and puts them next to acid rain as emblematic of the 1980s. It analyses the way in which the politics of environmental discourse influences the institutional repercussions of ecological modernization. The book ends with a discussion of how institutional reflexivity could be enhanced. On the basis of the empirical findings it argues for an appreciation of the merits of a public domain to accommodate the social conflict over the ecological dilemma. It discusses several possible institutional innovations that would enhance the reflexivity of our way of dealing with the ecological dilemma and concludes with two concrete suggestions: the introduction of a 'societal inquiry' and the system of 'discursive law'. It is argued that a strong public domain is the appropriate response to an anti-realist understanding of the environmental problematique. It should not be seen as a dysfunctional suggestion but as a way to enhance the institutional capacity to cope with the many different, and indeed often contradictory problem definitions that are characteristic for environmental discourse of modern society.

1

The New Environmental Conflict

1.1. THE ECOLOGICAL CRISIS AS DISCURSIVE CONSTRUCTION

If there is one image that has dominated environmental politics over the last twenty-five years it is the photo of the planet Earth from outer space. This picture, which entered the public imagination as an offspring of the 1960s Apollo space programme, is said to have caused a fundamental shift in thinking about the relationship between man and nature. The confrontation with the planet as a colourful ball, partly disguised by flimsy clouds, and floating seemingly aimless in a sea of utter darkness, conveyed a general sense of fragility that made people aware of human dependence on nature. It facilitated an understanding of the intricate interrelatedness of the ecological processes on planet Earth. Indeed, the image, it is said, caused a cognitive elucidation through which the everyday experience of life in an industrialized world was given a different meaning.

In the early 1970s the image of the globe became the icon of a comprehensive political effort to address global environmental problems and save 'Our Common Planet.'[1] The United Nations Conference on the Human Environment held in Stockholm in June 1972 had as its theme Only One Earth, and both the report *Limits to Growth* (1972) and the alternative *Blueprint for Survival* (1972) drew upon the image of the world as a 'biosphere', as one interacting whole.[2] The UN report *Our Common Future* (1987), which

[1] On the social construction and socio-political importance of this image, see also Sachs 1992: 107–9.

[2] Cf. Ward and Dubos 1972; Meadows *et al.* 1972; *Ecologist* 1972: 26. The expression 'Only One Earth' was actually borrowed from the peace movement that emerged in response to the new reality of the atomic age in the 1940s. See *One World or None: A Report to the Public on the Full Meaning of the Atomic Bomb*, with essays by Einstein, Bohr, Oppenheimer, and Walter Lippmann and others (Masters and Way, 1946).

became the centre-piece of the environmental debate in the late 1980s and laid the conceptual foundations for environmental politics in the 1990s, even suggested that the confrontation with the image of the planet had created a new reality, and suggested that the image might, eventually, be found to have had a greater impact on thought than the Copernican revolution. And here the image led to a call for action: 'This new reality, from which there is no escape, must be recognized—and managed.'[3] This appeal did not fall on deaf ears. Following the general endorsement of the Brundtland Report *Our Common Future* many Western countries published comprehensive documents outlining national environmental policy plans from around 1990. For all their differences, recent White Papers like the Dutch *Nationaal Milieubeleidsplan* (National Environmental Policy Plan—NMP, 1989) and *This Common Inheritance* (1990) in Britain both start from the recognition that the state of the environment calls for an integrated approach and outline a national strategy of bureaucratic regulatory management of the environmental problem, carefully positioning themselves within the context of the perspective of 'sustainable development' as proposed by *Our Common Future*. It is interesting to note that in both documents the photographic image of the planet earth as taken from outer space features prominently.[4]

One might, of course, argue that too much attention to one simple picture is misplaced.[5] Yet the career of the picture illustrates a profoundly disturbing feature of contemporary environmental

---

[3] WCED 1987: 1.

[4] The cover of the Dutch *Nationaal Milieubeleidsplan* shows the planet under a bold print of its sub-title: *Choose or Lose*. If one examines the photo more closely, it becomes obvious that the photo is actually manipulated. The image of fragility is given an extra dimension, since the surface of the Earth shows a crack that runs from pole to pole in what shows to be the very thin shell of the planet (cf. *Handelingen Tweede Kamer* 1988–9, 21 137, No. 1–2, and the sequence report, NMP-Plus in *Handelingen Tweede Kamer* 1989–90, 21 137, No. 20–1). The British report *This Common Inheritance* (DoE, 1990) shows the photographic image of the planet on the title page above the quote from John Stuart Mill from which the title is derived. By the first progress report the photographic image has been promoted to the cover (DoE 1991).

[5] The political meaning of the metaphor of the lonely planet has of course been pointed out before. The more explicitly ideological application of the image, such as in Kenneth Boulding's metaphor of the 'spaceship earth' or Hardin's 'life-boat' metaphor, have been widely criticized (cf. Enzensberger 1973: 18 ff.; O'Riordan, 1983: 27 ff. and 102). Kwa has rightly pointed out that the scientific image of the earth as a closed eco-system in fact has very much the same bias (Kwa 1987).

politics. The first naturalistic NASA photos may still be said to have been a straightforward eye-opener and catalyst for prudence: we take too much for granted and are ruining the things we value. And hence we might want to start to think about how to act upon this new reality: how can we act locally while thinking globally? Yet this straightforwardness is certainly not a feature of the latest representation of the planet from outer space: the computerized satellite picture of ozone depletion. Problems like ozone depletion or indeed the greenhouse effect seem to lack a material reality. We do not experience them; they are pointed out to us by experts that use high-tech devices such as colourful computer graphics to facilitate our understanding of these global threats. Their message is apocalyptic and their call is for major social change. The latest images of the planet leave little choice: it is choose or lose.

The environmental debates of the 1990s focus on problems that are much less the object of direct sensory perception (like smoke, dirt, or smells) or common sense understandings. They are different in terms of scale, time, and techniques. They take what has become known as the 'global biosphere' as their level of analysis, and portray problems that will often materialize many years from now. At the same time understanding has ceased to be a matter of direct experience, but is a matter of complex scientific extrapolations, of mathematical calculations that require extremely expensive supercomputers, and, consequently, it is a limited group of experts who define the key problems, who assess the urgency of one problem *vis-à-vis* other possible problems, and who implicitly often conceptualize the solutions to the problems they put forward. The layman, depending on sensory perception and everyday experience, is totally disqualified. He or she has become dependent and can only be educated. This disqualification in fact not only affects the proverbial man in the street: specialist natural scientists, politicians, philosophers, or social scientists, all experience how their stocks of knowledge and normative theories about proper procedural rules of reaching social agreements are devalued too.

The fact that smart scientists find complicated problems, try to explain them, and subsequently call for action is in itself, of course, no justification for concern. Yet there are at least two features of present environmental politics that could indeed be causes for concern: first, the changing basis of legitimate decision-making; and second, the hidden link between science and politics. First, as

Jerry Ravetz has pointed out, environmental experts call for extremely hard decisions but have only 'soft' evidence to support their claims. In 1972 people might have been shocked as *Limits to Growth* pointed out that economic prosperity could not be assumed to continue to grow indefinitely. Yet it is evident that the nature of environmental politics has changed considerably since then. People are now confronted with a wide range of views, with experts and counter-experts, with debates among scientists from different disciplines or different countries, and realize that scientific controversy is an inherent element of environmental politics. Hard decision-making on global environmental problems requires an almost unprecedented degree of trust in experts and in our political élites at the same time as this trust is continually undermined by scientific controversies and political indecision.

Second, implicitly global environmental problems are presented as being a priori of a different order, and thus marginalize many other environmental concerns that might affect many people or eco-systems much more directly.[6] The apocalyptic overtones in the presentation of global environmental problems seriously confines the political debate on what needs to be done, by whom, and under what conditions. Hence there is ample reason to examine in detail how the priority issues in environmental politics are defined, as well as trying to understand the implicit political decisions that are being made through environmental discourse.

The way in which environmental politics is made is an intricate matter and it is unproductive to think that there is a short-cut. It seems wrong, for instance, to suggest that science is merely used as a fig-leaf for policies. Likewise it seems a mistake to think that there is a clear-cut coalition of actors who construct these global environmental problems to further their own preconceived goals. In fact the globalization of environmental discourse seems to be much more the product of a very varied set of actions that have all in some way contributed to the present dominance of global environmental problems. Nevertheless, we can observe how the Brundtland Report functioned as the catalyst for change in environmental policy. There it is argued that

---

[6] For instance, largely localized environmental problems such as industrial toxic pollution, desertification, or soil erosion. As Buttel and Taylor (1992: 218) have pointed out, these problems are not global but localized in their antecedents, social consequences, and environmental implications.

The time has come to break out of past patterns. Attempts to maintain social and ecological stability through old approaches to development and environmental protection will increase instability.... We are unanimous in our conviction that the security, well-being, and very survival of the planet depend on such changes, now.[7]

The Report was effective in proliferating an image according to which the twenty years that passed since 1972 are to be understood as a period of progressive social learning and an enlightened institutional crusade. Slowly awareness grew and political and organizational resistance gave way to a broad consensus around the notion of 'sustainable development.'[8] Now time has come for serious policy-making.

It is undoubtedly true that such an institutional crusade has taken place. Indeed, the Brundtland Report was part of it. One of the main achievements of the Brundtland Report is to have presented the environmental case in such a way that it could bring round big institutions like the World Bank and the IMF which, in the 1970s, were still considered to be in the opposing camp. But this is where the potential problems with the Brundtland approach lie. Radical critics of the Brundtland Report claim that the whole idea of sustainable development is a rhetorical ploy which conceals a strategy for sustaining development rather than addressing the causes of the ecological crisis.[9] Indeed, after the Earth Summit, held in Rio de Janeiro in June 1992, had made the World Bank one of the central agencies monitoring the greening of policies all over the world, radical critics argued that this did not prove the success of the environmentalists, but their total collapse.[10]

This discussion illustrates that people may have widely differing perceptions of what environmental politics is about. In this light the present hegemony of the idea of sustainable development in environmental discourse should not be seen as the product of a linear, progressive, and value-free process of convincing actors of the importance of the Green case. It is much more a struggle between various unconventional political coalitions, each made up of such actors as scientists, politicians, activists, or organizations

[7] WCED 1987: 22–3.
[8] For good discussions of the concept of sustainable development and the Brundtland Report, see Clark and Munn 1986; Redclift 1987; de la Court 1990; Pearce *et al.* 1989.
[9] Sachs 1992.   [10] See Finger, 1993; Shiva 1993.

representing such actors, but also having links with specific television channels, journals and newspapers, or even celebrities. These so called discourse coalitions somehow develop and sustain a particular discourse, a particular way of talking and thinking about environmental politics.[11] These coalitions are unconventional in the sense that the actors have not necessarily met, let alone that they follow a carefully laid out and agreed upon strategy. What unites these coalitions and what gives them their political power is the fact that its actors group around specific story-lines that they employ whilst engaging in environmental politics. It can be shown that although these actors might share a specific set of story-lines, they might nevertheless interpret the meaning of these story-lines rather differently and might each have their own particular interests. Take the story-line of the rainforests as an example. The systems-ecologist might insist on the importance of the rainforests as an essential element in his or her mathematical equations that model the world as a biosphere, as an integrated and self-reproducing eco-system; the World Wildlife Fund is more concerned about the moral problem of forest destruction; while the singer Sting connects the fate of the rainforest to that of the culture of the indigenous people, thus stretching the idea of habitat protection to its limits. NASA may add to the credibility of the story-line through the publication of satellite photography showing the change of forest cover over time.

To be sure, there can be no doubt about the fact that the rainforest cover is in decline, but what exactly is the problem? All actors speak about the rainforest but mean (slightly) different things. If examined closely, the various actors have rather different social and cognitive commitments, but they all help to sustain, in their own particular way, the story-line of the destruction of the rainforests in environmental politics. Once the story-line gets enough socio-political resonance it starts to generate political effects, but who then controls the meaning of the rainforest story-line? What happens in the discursive construction of the rainforest as a public problem? How do the discursive construction and institutional response influence one another?

This book argues that the environmental conflict has changed. It has become discursive. It no longer focuses on the question of

---

[11] The concept of 'discourse-coalition' is expanded in Ch. 2.

whether there is an environmental crisis, it is essentially about its interpretation. 'We are all Greens now', politicians can be heard saying. Talking Green no longer connotes a radical social critique, and here the image of the planet evokes sentiments that help to create and sustain a perception of a common global ecological crisis, which implies shared values and common interests. As such it functions as a symbolic umbrella, as an inclusionary device, that constitutes actors as joint members of a new and all inclusive 'risk community'. The image of the planet, once the catalyst for reconsidering our local environmental problems, has become the symbol of the view that the real dangers are those of a global physical crisis that threatens survival. According to this story-line the environment has become a problem of mankind which can only be resolved by one big united effort. As such the employment of the image of the planet is as much disempowering to more situated or indeed social understandings of environmental problems and its solutions as it is instrumental to the formation of a political consensus on the need for comprehensive and centralized global action.

What is forgotten is the issue of representation. As Benton and Redclift rightly ask, 'Do we share an understanding of the global environment in the same way as we "share" the globe?'[12] It is my contention that this is not the case. To the extent that we do have a similar understanding of the global environment and its problems this is the product of the representations according to which we understand environmental change. Hence sustainable development should also be analysed as a story-line that has made it possible to create the first global discourse-coalition in environmental politics. A coalition that shares a way of talking about environmental matters but includes members with widely differing social and cognitive commitments. The paradox is that this coalition for sustainable development can only be kept together by virtue of its rather vague story-lines at the same time as it asks for radical social change.

This book examines the nature of the new environmental conflict. It argues that the new environmental conflict should not be conceptualized as a conflict over a predefined unequivocal problem with competing actors pro and con, but is to be seen as a complex and continuous struggle over the definition and the

---

[12] Benton and Redclift 1994: 16.

meaning of the environmental problem itself. Environmental politics is only partially a matter of whether or not to act, it has increasingly become a conflict of interpretation in which a complex set of actors can be seen to participate in a debate in which the terms of environmental discourse are set. The method used is discourse analysis, which is employed to illuminate the social and cognitive basis of the way in which problems are constructed. For this purpose the book examines the interaction between the social processes through which actors are mobilized around certain issues with the specific ideas and concepts that create common understandings of given problems.

This approach has major repercussions for the spatial focus of political analysis. I will argue that if examined closely, environmental discourse turns out to be essentially fragmented and contradictory. This is not surprising since environmental discourse is in fact the product of the interaction that takes place in practices that often lie well beyond the traditional political realm. Those discursive elements will often contradict one another, but at least as often concern quite separate aspects of reality. My interest in this book is to illuminate how certain dominant perceptions of a problem are constructed and how political decision-making takes place in this context of, and through, essentially fragmented and contradictory discourses within and outside the environmental domain.

Given the theoretical approach that draws on the work of Foucault and social psychologists like Harré and Billig, I consider it essential to look in detail at the specific practices through which common understandings are produced and transformed. The focus of this book is on the way in which these discursive elements are then mediated and drawn upon in policy discourses. This also puts policy-making in a different perspective. Policy-making is in fact to be analysed as the creation of problems, that is to say, policy-making can be analysed as a set of practices that are meant to process fragmented and contradictory statements to be able to create the sort of problems that institutions can handle and for which solutions can be found. Hence policies are not only devised to solve problems, problems also have to be devised to be able to create policies. The book aims to bring out the extent to which the practices of policy-making succeed in regulating social conflicts over environmental matters.

In Section 1.2 I will present some basic premises on the basis of which the argument will proceed. Section 1.3 introduces the way in which I will go about analysing environmental regulation. Section 1.4 presents an outline of the new policy discourse of ecological modernization that now dominates the thinking about what environmental policies should be about. Section 1.5 introduces the normative questions against which we want to judge the way in which ecological modernization affects the regulation of key environmental problems. Finally, Section 1.6 presents our questions for research and introduces acid rain as an appropriate case for the study of the effects of the discourse of ecological modernization on contemporary environmental politics.

1.2. THE STATUS OF NATURE IN ENVIRONMENTAL DISCOURSE

This book takes issue with those analyses that start from a realist definition of the ecological problem and concentrate their analytical work on explaining why action is slow in coming, and who or what determined whether action was taken. A realist approach assumes incorrectly that the natural environment that is discussed in environmental politics is equivalent to the environment 'out there'. This assumption fails to recognize that we always act upon our images of reality and are dependent on certain discourses to be able to express ourselves. There images and discourses should themselves be included in our analysis. Cultural anthropologists have illuminated the cultural differences in our perception of nature. One culture may conceive nature as fragile and live with the certainty that natural equilibria are easily disrupted, another culture may be convinced that nature is robust and can therefore cope with certain disturbances.[13] Actors may have quite distinct ideas of what the ecological crisis is about, with ideas about what sort of action should be taken differing accordingly.

Behind this is the epistemological problem of how we can know what nature really is. The intricacies of the philosophical debate go beyond the scope of this book, yet we can and should infer an essential feature from it. Both the Kantian distinction between the

---

[13] Cf. e.g. Thompson *et al.* 1990; Douglas 1992.

Ding-an-sich and our conceptions (*Vorstellung*) of the *Ding-an-sich*, and the Lacanian differentiation between the 'Real' and 'reality' seek to clarify the knowledge of the subject about the world 'out there'. Both distinctions are based on the idea that what we know is being framed by experiences, by languages, by images, or even by human fantasies. Following Lacan we see all our knowledge of nature (and indeed, society!) as essentially metaphorical. We set apart the Real as a 'hard kernel, which resists any process of modelling, simulation, or metaphoricization.'[14] Reality, then, is always particular, it is always dependent on subject-specific framing or time-and-place specific discourses that guide our perceptions of what is the case.

This should not be misunderstood as an argument that denies the existence of severe ecological problems. It does, however, seek to qualify our statements on the state of the environment. As Neil Everden put it, 'We must bear in mind that the current understanding of pollution is just that: the current understanding.'[15] Environmental discourse is time- and space-specific and is governed by a specific modelling of nature, which reflects our past experience and present preoccupations. Any understanding of the state of the natural (or indeed the social) environment is based on representations, and always implies a set of assumptions and (implicit) social choices that are mediated through an ensemble of specific discursive practices. This does not mean that nature 'out there' is totally irrelevant. Yet the essence of the argument is that the dynamics of environmental politics cannot be understood without taking apart the discursive practices that guide our perception of reality. In this section I will present five fundamental points that underscore the importance of this social constructivist orientation on the discursive practices in which environmental politics is made.

First, geographers working in the tradition of George Perkins Marsh have documented that environmental change cannot be seen as a temporary phenomenon, but is structural in character.[16] Mankind has always used nature to further its own goals and this manipulation of the natural environment has been accompanied by a range of serious environmental problems. In this sense it is

---

[14] Cf. Žižek 1993: 44.
[15] Cf. Everden 1992: 4. More on the theoretical argument follows in Ch. 2.
[16] Cf. Glacken 1967; Marsh 1864; Leiss 1978; Thomas 1956; Simmons 1989.

appropriate to refrain from using the label environmental 'crisis' and speak of the environmental 'dilemma' of industrial society instead. We can, however, differentiate between the features of environmental change that are problematized in distinct periods and try to explain these conflicts out of the interaction between physical change, changing social practices, and specific social sensibilities. In that case we in fact no longer analyse ecological problems but socio-ecological problems.

Second, debates on pollution always raise questions about the social order in which the pollution occurred. In her comparative study *Purity and Danger* the anthropologist Mary Douglas defined dirt or pollution as 'matter out of place.'[17] This succinct definition presents a very astute understanding of what debates on pollution are about. Environmental change is of all times and all societies but the meaning we give to physical phenomena is dependent on our specific cultural preoccupations. Douglas shows that what is seen as 'out of place' depends on the social order within which meaning is given to phenomena.[18] We can infer that attempts to redefine what is seen as dirt or pollution are implicit challenges to the social order since, as Douglas puts it, 'disorder spoils pattern.'[19] To analyse discourses on pollution as quasi-technical decision-making on well-defined physical issues thus misses the essentially social questions that are implicated in these debates. Alternatively, the study of environmental discourse can bring out how socio-ecological questions of environment and social development are dealt with politically.

Third, debates on nature or pollution reflect the contradictions of the social developments against which they take place and of which they are part. This point is well made in the brilliant study *Man and the Natural World*, by the historian Keith Thomas. He describes the changing attitudes of the British people to the natural environment between 1500 and 1800 and illuminates how, in a phase of growing industrialization and urbanization, the growing

---

[17] Douglas 1988: 35.

[18] This social element almost unnoticeably slips into many more strictly scientific definitions of pollution, for instance as given by Holdgate: 'The introduction by man into the environment of substances or energy liable to cause hazards to human health, harm to living resources and ecological systems, damage to structures or amenity, or interference with *legitimate uses* of the environment.' (Holdgate 1979: 17, emphasis added).

[19] Douglas 1988: 94.

domination and control of man over nature changed the attitudes of the British towards nature. It leads him to the conclusion that the relation of man and nature is governed by what he calls a 'human dilemma'. The question seems to be

how to reconcile the physical requirements of civilisation with the new feelings and values which the same civilisation had generated . . . the growth of towns had led to a new longing for the countryside. The progress of cultivation had fostered a taste for weeds, mountains and unsubdued nature. The new-found security from wild animals had generated an increasing concern to protect birds and preserve wild creatures in their natural state.[20]

This paradoxical state of affairs is certainly still with us today. In Thomas's account the paradox is resolved in the suggestion that the discourse on nature was a somewhat ambivalent, romantic reaction to societal modernization that celebrates nature and beauty as opposites to the harsh everyday reality of trade and industry. On this point Raymond Williams has argued that environmentalism mostly failed to address the social relations of which the very predicament was the result.[21] This ambivalence in environmental discourse is also a fundamental point in Hays's celebrated study on the US conservation movement.[22] Hence debates on nature or pollution may reflect the contradictions of the social developments against which they take place but it is an empirical question whether or not (or how) these developments are problematized or how these ambivalences are being managed. More particularly, political analysis should illuminate the places, moments, and institutions where certain perceptions of environmental change and social development emerge and are reproduced, and should reconstruct the argumentative struggle that determines which perceptions at some point start to dominate the course of affairs in environmental politics.

Fourth, the environmental problematique is hardly ever discussed in its full complexity ('in the round'). Environmental discourse tends to be dominated by specific emblems: issues that dominate

---

[20] Thomas 1983: 301.

[21] Williams 1985. Obvious exceptions in 19th-cent. Britain were social critics like John Ruskin and William Morris, who combined the quest for preservation with explicit social criticism and came up with an argument for the redistribution of wealth and reorganization of society.

[22] Hays 1979.

the perception of the ecological dilemma in a specific period. To argue that certain issues can be emblematic obviously is not meant to suggest that these issues are irrelevant or draw away attention from the 'real' issues.[23] The political importance of emblems in environmental discourse is that they mobilize biases in and out of the environmental debate. They are the issues in terms of which people understand the larger whole of the environmental condition. As such, they thus effectively function as a metaphor or, to be more precise, as a metonym.[24] Examples of emblems in environmental discourse of the last 150 years include deforestation in the nineteenth century, the destruction of the countryside (UK) or wilderness (US) around the turn of the century, soil erosion in the 1930s, pesticide pollution in the early 1960s, resource depletion in the early 1970s, nuclear power in the late 1970s, and global issues like the greenhouse effect and the diminishing ozone layer in the 1980s.[25] The task of political analysis is to look at how actors are mobilized around such emblems and to examine the implications of this process of 'coalition formation' for the environmental discourse. Referring back to what was said in connection with Douglas above, the study of environmental discourse should illuminate which questions about social developments and which social expectations can be discussed meaningfully in the context of these emblems.

Fifth, discursive strategies matter. Today's environmental issues are discursively created. A leaking oil tanker, for example, is of course a physical event in itself, but then so is an unreported chemical spillage. Calamities only become a political issue if they are constituted as such in environmental discourse, if story-lines

[23] However, many of the biggest environmental disasters might well be extremely slow in becoming apparent or do not show as disasters at all as, for instance, with increases in cancer incidence or the repercussions of shortages of safe drinking water (see Scimemi 1988: 36 ff.). In this sense the charismatic issues that become emblems may indeed draw attention away from issues with greater environmental or health impacts.

[24] A metonym is a technical linguistic term for mistaking a part for a whole, as in saying 'I would like to have a Coke please' when one intends to order a cola (of whatever brand); cf. Lakoff and Johnson 1980: 35 ff.

[25] For deforestation in the 19th cent., see Marsh, 1864; for emblems around the turn of the century, see Hays 1979; for soil erosion in the 1930s, see Worster, 1991; for pesticide pollution in the early 1960s, see Carson 1963; for resource depletion in the early 1970s, see Meadows *et al.* 1978; for nuclear power in the late 1970s, see Cotgrove and Duff 1980; for global issues in the 1980s, see WCED 1987.

are created around them that indicate the significance of the physical events (compare, for in stance, the effect of 'It is the fourth
consecutive spillage in two months' or 'The tanker did not have
doubly secured partitions' with 'The stormy weather will guarantee the chemical breakdown of the oil'). Here they depend on
agency and discursive strategies. This is even more true for a
phenomenon like acid precipitation (quickly redefined as 'acid rain')
or global warming (discussed as the 'greenhouse effect'), which are
not clearly identifiable 'events' but are rather gradual processes,
accidents in slow motion, as Roqueplo has put it.[26] Here the
publication of a 'report on survey findings' or 'new computer
extrapolations' replace the reality of an invisible calamity while
press conferences are the second intermediary practice needed to
make an environmental hazard into a political issue.[27] Furthermore, the dominant role of emblems in environmental discourse
indicates that single issues determine the public perception of a
much more complex reality: hence if one (type of) disaster receives
attention, other issues might pass relatively unnoticed. The discursive construction of reality thus becomes an important realm of
power. If a discourse of concern could be diverted away from
certain aspects of reality, no further action is needed. However, no
one obviously creates a calamity single-handedly (or steers one
away, for that matter). Discursive strategies have to be understood
in their own social and cognitive context.

1.3. REGULATING THE ENVIRONMENTAL CONFLICT

What I am interested in is the way in which social conflicts over
environmental emblems are handled politically. This is done by
examining the regulation of these issues by means of environmental policy-making. First we will deal with the way in which we
should understand the term 'policy-making' and subsequently we
will describe how 'regulation' is analysed.

In a constructivist approach policy-making is not seen simply as

---

[26] See Roqueplo 1986.
[27] If we leave out for the moment other practices, for instance those related to
the role of the media in deciding whether something constitutes 'news', or the the
practice of NGOs of selecting certain Campaign issues on which they concentrate
their activities for a period of time.

a matter of what I call 'problem closure': defining a set of socially acceptable solutions for well-defined problems. It is, first and foremost, an interpretive activity in which different, and often contradictory claims as to what is the case are to be judged, compared, combined, and acted upon. The definition of a policy problem, like for instance acid rain, cannot be taken for granted. Acid rain is in fact made up of 'an historically constituted set of claims', to use Forester's words.[28] Experts make claims with regard to the state of the environment; other actors promote specific solutions; politicians try to maintain an image of being in control of a situation; while NGOs may frame acid rain in terms of a preconceived critique of existing institutional arrangements; yet others may introduce the ethics of good-neighbourliness; while pollution inspectorates may deny certain accusations just to conceal their structural regulatory failure. These various constitutive elements of the problem of acid rain are nearly all contested. Yet at the same time one can see how out of these discursive fragments certain claims 'somehow' become related to one another and result in a particular definition of the policy problem. This process may be called the 'discursive closure' of policy problems.

There is a second addition to the understanding of policy-making as problem closure (which derives its formulation from the discipline of public administration). In a social scientific analysis policy-making also appears as the dominant form of regulation of social conflicts in modern industrial societies. Latent social conflicts that might erupt over environmental change are predominantly accommodated through this subtle discursive process rather than through the use of force or manipulation.[29] In light of the above, I distinguish three main tasks of regulation. The regulation of a problem first and foremost requires forms of discursive closure: the problem needs a definition that gives policy-making a proper target. Secondly, regulation asks for social accommodation: policy-making implies finding ways to contain the social conflict that might erupt over environmental problems. Thirdly, regulation is supposed to allow for problem closure: it should

---

[28] Forester 1982: 43.

[29] Hence in this book regulation directly refers to conflict management and differs from other understandings of regulation as enforcement and implementation (Hawkins 1984; Vogel 1986) or as an alternative to welfare policy (Majone 1989a, 1989b).

remedy a situation that was perceived as problematic. This perhaps unusual definition serves the following purposes. At the level of discursive closure we can analyse what aspects of the problem are included and what is left out. The analysis of the social accommodation adds the understanding of the particular way in which problems are positioned in the contexts of some social developments and aspirations while others are deflected. The analysis of problem closure shows whether the strategy of regulation helped to achieve certain environmental goals.⟩

It will be obvious that these three main tasks are not necessarily mutually supportive and are, potentially, even contradictory. Solutions that are invented to deal with a purely technically defined physical problem but that are insensitive to the common-sense social construction of a problem are likely to become regulatory failures.[30] What is more, the literature on the social perception of risk has amply shown that fears and worries of citizens are likely to persist if no attempt is made to reconcile the technical perception of the problem with the social dimension.[31] Conversely, if interventions (or non-interventions for that matter) simply follow prevalent social constructs of a particular problem, it may perhaps secure considerable social support but may aggravate the state of the physical environment or, more precisely, fail to come to problem closure for the problem as described in expert discourse.

Section 1.2 has made clear that environmental discourse should not be approached as a coherent whole but is better seen as inherently contradictory and ambivalent. In this context the art of regulation is to find a way to secure credibility in face of these contradictions, to render regulatory strategies acceptable, and to generate trust for the institutions that are put in charge of regulation. Here discursive strategies constitute an essential variable for political action both of policy-makers[32] and their critics, since the image of problems that can be regulated (or have to be regulated), are defined through the employment of specific historical references, symbols, metaphors, etc. What is more, in this process of setting the terms of a policy discourse credibility and authority are allocated, for instance by emphasizing a tradition of effective policy-making in a particular field or precisely by dissociating a

---

[30] See Forester 1982: 52–3.   [31] Wynne 1982; Fischhoff *et al.* 1981.
[32] Policy-makers are defined as responsible politicians and their senior administrative advisers.

'new' problem from previous ones, if such a tradition could not be posed credibly. Regulatory failure, in other words, is something that can only be established after a careful analysis of the different perceptions of a particular problem.

If key policy problems like acid rain, global warming, or the diminishing ozone layer, are to be seen as examples of historically constituted sets of claims, the question for empirical research is what these claims were, where these claims came from, and above all, how they were somehow combined and recognized as a policy problem. We will rest this case here and take it up again in Chapter 2, which is devoted to the theoretical argument about how coalitions in a specific political domain uphold certain structures through which meaning is allocated. An essential part of that argument is that the dynamics of emblems should be understood as a constant interaction between the development of a general frame or policy discourse within which environmental problems are to be understood on the one hand, and the development of insights that are directly related to the actual policy problem on the other. It is my contention in this respect that present-day environmental controversies should be positioned against the background of an important change in these general policy discourses that has occurred in the Western world since 1972.

1.4. THE AGE OF ECOLOGICAL MODERNIZATION

1972 is often taken as the starting-point for the wave of environmental politics. It was the year of the publication of the report to the Club of Rome *Limits to Growth* and it saw the UN conference of the Environment, held in Stockholm, Sweden. This was at its time the biggest UN conference ever held. The 'environmental problematique', as the Club of Rome called it, has been on the agenda ever since, although the focus of debate has been on differing emblems. Yet in many regards environmental politics has also changed considerably since that time. What interests us here are the dramatic changes in the way in which environmental policies are conceptualized. Immediately following the ascent of the environmental problematique in the hierarchy of political attention in the early 1970s, most Western countries created the environment as a (semi-)independent field of attention for the first time.

Ministries were set up, though they were often attached to the established health departments. Characteristic of this early period was the predominantly legislative attempt to try to cover all aspects of environmental reality with at least a basic set of rules on legitimate conduct concerning the emission of substances. This took the classic bureaucratic form of functional differentiation, dividing reality into smaller sections drawing on the so-called 'compartmental division' dividing the environment into air, water, soil, and—sometimes—sound. Organizational structures and legislation reflected the same division. Essential to the first institutional response was the fact that pollution as such was not regarded as problematic: in most cases the basic concern was to guarantee certain general quality norms. Environmental policy was a matter of aiming to control the quality of these separate compartments by imposing overall quality targets or regulating various implicated industrial works through the handing out of pollution permits. Characteristic of this era was the basically subservient role of environmental protection to industrial politics. This was not too surprising since pollution was not generally recognized as a structural problem: it was basically perceived to be a problem that could be contained using *ad hoc*, and *ex post* remedial measures. Typical of the abatement strategies that were applied under this regulatory regime were 'end-of-pipe technologies' such as filters on chimneys or drains, or collectively financed water-processing plants.

This type of environmental policy discourse lost much ground in governmental circles between 1972 and 1990. Indeed, the 1980s saw the emergence of a new policy-oriented discourse in environmental politics that could be labelled 'ecological modernization'.[33] In the most general terms ecological modernization can be defined as the discourse that recognizes the structural character of the environmental problematique but none the less assumes that existing political, economic, and social institutions can internalize the care for the environment. For this purpose ecological modernization, first and foremost, introduces concepts that make issues of

---

[33] The concept of ecological modernization was first introduced by the German political scientists Joseph Huber and Martin Jänicke (Huber 1982, 1985, 1991; Jänicke 1985, 1988). These authors allocate a central role for technological innovation and economic development in ecological modernization. As I will show, the conceptual change in fact stretches to many other domains (see also Spaargaren and Mol, 1992). For the social history of eco-modernist discourse, see Ch. 3.

environmental degradation calculable. Most notably, ecological modernization frames environmental problems combining monetary units with discursive elements derived from the natural sciences. This provides a common denominator through which costs and benefits of pollution can be taken into account. A second characteristic is the fact that environmental protection is portrayed as a 'positive-sum game'. Likewise, the main obstacles to more effective protection are suggested to be dilemmas of collective action: there would be no fundamental obstructions to an environmentally sound organization of society, if only every individual, firm, or country, would participate. Environmental protection thus becomes a management problem. A third and related characteristic is the fundamental assumption that economic growth and the resolution of the ecological problems can, in principle, be reconciled. Hence, although some supporters may individually start from moral premises, ecological modernization basically follows a utilitarian logic: at the core of ecological modernization is the idea that pollution prevention pays. The 1987 Brundtland Report *Our Common Future* can be seen as one of the paradigm statements of ecological modernization. The shift to the discourse of ecological modernization represents a general trend in the Western world. That is to say, we can see the same ideas, concepts, divisions, and classifications emerging in different countries and international organizations, such as the UN, the OECD, or the European Union.

The conceptual shift to ecological modernization can be observed in at least six different realms. The primary sphere where the shift to ecological modernization occurs is in the techniques of environmental policy-making. During the 1980s environmental policy-making came to be organized around a wholly new set of principles. Here at least two tendencies intermingle. First, the traditional judicial administrative structures that had been put in place in the 1970s were problematized. The deficiencies of the predominantly 'react-and-cure' formula for regulation were increasingly criticized, while the more innovative 'anticipate-and-prevent' variety gained credibility. Cross-boundary pollution illuminated further functional deficiencies. The functional frustrations with the compartmental division led to a call for an integrated approach to pollution abatement, and, likewise, the integration of pollution prevention into the activities of other ministries received more attention. A second tendency is at least as important. The hierarchical legislative system often involved complicated administrative procedures which became problematic in a period in which deregulation became widely accepted as one of the goals of government. Accordingly, many new techniques were introduced that were supposed to allow individual firms to integrate environmental concerns into their overall calculation of costs and risk. Over the last two decades we thus saw the introduction of, more or less in order of appearance, the polluter pays principle, cost–benefit analysis, risk analysis, the precautionary principle, tradeable pollution rights and the levy of charges on polluting activities, as well as the debate on resource taxes and emission taxes.[34]

Secondly, ecological modernization shows in a new role for science in environmental policy-making. In the first instance science had as a prime task to come up with proof of damaging effects and was thus active on the policy input side. The changing social perception of environmental change (which was partly the result of pioneering work by individual scientists) meant that science increasingly became entangled in the centre of the process of policy-making. Here the science of ecology, and especially systems ecology, starts to fulfil an increasingly important role. This is manifest in the emergence of concepts like 'multiple stress' and 'critical load'. The increased credibility of the first concept signifies a shift from a reductionist scientific ontology and epistemology towards an orientation that takes more integrative ecological ideas about nature as its starting-point. The concept of critical level was imported from the field of health policy and was redefined as 'critical load'. The concept of critical load precisely serves the applied purpose of establishing the uptake capacity of natural ecosystems. The scientist is thus given the task of determining the levels of pollution which nature can endure.

Thirdly, on the micro-economic level the shift to ecological modernization surfaces in the move away from the idea that environmental protection solely increases cost to the concept of 'pollution prevention pays'. This idea, that has long intellectual roots,

---

[34] It should be observed that some countries were more reluctant to apply these concepts of ecological modernization. Yet even in the United Kingdom, which, apart from one particular period in the mid-1980s, kept a rather reserved approach to the introduction of new principles, concepts like the polluter pays, cost–benefit analysis, or the levy of charges were either accepted by the government in the 1970s or were at least under discussion in respected advisory councils like the Royal Commission on Environmental Pollution: see Royal Commission 1971, 1984.

was taken up following the resource crisis of the early 1970s and led to a number of influential publications.[35] It was pioneered in the 1970s by American firms like 3M, but has really penetrated management practice in America and Europe from the mid-1980s onwards. On the policy side this led to the promotion of 'low and non-waste technologies', and gave rise to the idea of 'multi-value auditing' (measuring success not only in terms of money but also taking energy and resource usage into account). Anticipatory investments thus replaced the basically unproductive investment in, for instance, 'end-of-pipe' technologies. Its leitmotiv is that if you don't put a substance into the environment, you don't have to pay to get it out.

Fourthly, on the macro-economic level eco-modernist thinking conceptualizes nature as a public good or resource instead of the idea that nature is basically a free good and can be used as a 'sink'. In this respect ecological modernization seeks to put an end to the externalization of economic costs to the environment or third parties. Furthermore, it puts great emphasis on the need to conserve or manage the scarce natural resources,[36] stimulating ecological pricing, recycling, and technological innovation.[37]

Fifthly, the legislative discourse in environmental politics also changes character. Given the fact that nature can no longer be regarded as a sink and that firms are now increasingly supposed to prevent pollution, the strict proof of causality by the damaged party often makes way for the idea that the burden of proof should be the concern of the suspected individual polluter, not the damaged or prosecuting party. In this discursive frame statistical probability and correlation became the basis for collective and unlimited liability.[38]

Sixthly and finally, ecological modernization implies a reconsideration of the existing participatory practices. Ecological modernization is based on the acceptance of the existence of a comprehensive environmental problem. This also means that it seeks to bring to an end the sharp antagonistic debates between the state and the environmental movement that were characteristic

---

[35] See Royston 1979; Huisingh and Bailey 1982; Huisingh *et al.* 1986.

[36] Here the old macro-economic ideal of maximizing flows is dropped in favour of a materials balance approach: see Leiss 1978; Kneese *et al.* 1971.

[37] See e.g. Daly 1977.   [38] See Teubner *et al.* 1994.

for the 1970s. Ecological modernization acknowledges new actors, in particular environmental organizations and to a lesser extent local residents. Hence ecological modernization also shows itself in an opening up of the existing policy-making practices and the creation of new participatory practices (from regular consultancy to active funding of NGOs, from a reconsideration of the procedural rules of Environmental Impact Assessments to the regular employment of round table discussions and environmental mediation).

A new policy discourse as comprehensive as ecological modernization is not conceptualized as one united set of ideas but only gradually emerges after yeas of institutional debate. Sometimes its emergence can only be recognized retrospectively, yet mostly there will have been a hard core of ideas to which other ideas responded only to create a more or less coherent shift in emphasis that may bridge several related domains. In this case the ideas of ecological modernization had been floating around in various spheres for some time.[39] From the late 1970s onwards we see story-lines emerging on the successful practices of 3M, Germany, or Japan. In Chapter 3 I will show how ecological modernization became the dominant discourse as a result of a specific argumentative interplay between governments, environmental movements, and key expert organizations. Here I confine myself to stating that in about 1984 ecological modernization was recognized as a potential alternative for policy-making by national governments and supra-national organizations in Europe and Japan. This is especially obvious in international environmental politics. The European Community's Third Action Plan for the Environment of February 1983[40] was written in eco-modernist spirit, and ecological modernization became the explicit focus of the discussion at the influential OECD conference on Environment and Economics of June 1984.[41] The ministers of the European Community agreed

[39] The need for an anticipatory strategy had been discussed in academic circles since the late 1960s and was also evident in the World Conservation Strategy (1980). As policy discourse it was promoted by the OECD who first introduced the theme in the council of ministers for the environment in May 1979: see OECD 1980: 5–10; OECD 1981a; see also Ch. 3.

[40] Commission of the European Communities 1983.

[41] See OECD 1985. The conference was attended by delegations from OECD member countries made up of ministers, MPs, and senior civil servants as well as representatives from the world of industry, trade unions, academia, and environmental NGOs.

on a statement that called for the 'wise management' of the resources of the environment and called for the 'integration' of environmental policy into other policies, especially economic decision-making. It called for the endorsement of 'sustainable development' and came out in favour of 'anticipatory environmental policies.'[42] At the subsequent Economic Summit held in Bonn on 4 May 1985, environmental protection and economic progress were put forward as necessary and mutually supportive goals. As was said above, between national governments there was a marked difference in the extent to which they actually acted upon these ideas, yet ecological modernization was most certainly part of the debate. Sometimes authoritative advisory councils introduced them, or because they came up in think-tanks (both of which were the case in the UK), sometimes a government opted to restructure a large part of its policy-making on this footing (as in Germany or Japan). It is these alternative ways of conceptualizing environmental problems that marks the debate on the regulation of emblematic issues like acid rain or defines the appropriate response of national governments to problems such as the greenhouse effect.

From the early 1980s onwards, ecological modernization increasingly came to dominate the debate on environmental regulation. Although the debate has taken many twists and turns over the years, and although ecological modernization has come to mean many different things to different actors in different places, it is still reasonable to claim that ecological modernization as outlined above has begun to dominate the conceptualization of problems, solutions, and the social strategies through which regulatory achievements are to be made.

My claim is not that ecological modernization is hegemonic in the sense that no other discourses are to be found in the environmental domain. The thesis is that over the second half of the 1980s ecological modernization has become the most credible way of 'talking Green' in spheres of environmental policy-making and increasingly functions as the organizing principle for the innovation of institutional procedures in the early 1990s. What we want to investigate are the political consequences of this change in environmental discourse.

---

[42] OECD 1984.

1.5. TOWARDS A CRITICAL APPRAISAL OF ECOLOGICAL MODERNIZATION

It is not difficult to see why the policy discourse of ecological modernization would appeal to governments. At least four reasons can be given. First of all, it positions itself in contradistinction to the *ex post* remedial strategy of the 1970s that did not produce satisfactory results. In face of that regulatory failure a new regime had to be devised. Ecological modernization can be read as a direct critique of these bureaucratic practices. The 1970s idea to control environmental pollution, dividing the environment into 'components' (air, water, soil, noise) and then drawing on specialized knowledge to define routine solutions for each sub-category, had failed. If anything, ecological modernization promotes cleverer regulatory mechanisms: fingers, not thumbs, to use Lindblom's phrase.

Secondly, ecological modernization suggests a positive-sum solution to what had until then been seen as a zero-sum problem. Governments are well aware of their functional dependency relationship with business and generally realize that calling a halt to environmental degradation would normally involve imposing restrictions on industry.[43] Ecological modernization, however, uses the language of business and conceptualizes environmental pollution as a matter of inefficiency, while operating within the boundaries of cost-effectiveness and administrative efficiency. Ecological modernization is the positive approach to environmental policy: environmental improvement does not have to be secured within the constraints of capitalist market logic (which would be a negative argument), ecological modernization suggests that the recognition of the ecological crisis actually constitutes a challenge for business. Not only does it open up new markets and create new

---

[43] A most succinct description of the mechanism which makes government performance depend on business performance is given by Lindblom: 'Depression, inflation and other economic distress can bring down a government' (see Lindblom 1977: 172–3). Hence governments that find themselves competing for capital investment in an increasingly open international economy (see Katzenstein 1985 but also Harvey 1989), must give business incentives or inducements to perform their task satisfactorily. Although the level of privilege may vary according to time and place, this mechanism implies that various fields of governmental policy, including environmental policy, are 'curbed and shaped by concern for possible adverse effects on business.' (Lindblom 1977: 178).

demands; if executed well, it would stimulate innovation in methods of production and transport, industrial organization, consumer goods, in short, all those elements that Schumpeter once identified as the forces that produce the 'fundamental impulse that sets and keeps the capitalist engine in motion.'[44] In this sense the discourse of ecological modernization puts the meaning of the ecological crisis upside-down: what first appeared a threat to the system now becomes a vehicle for its very innovation. Indeed, by now it is obvious that activities like waste management, pollution abatement, and risk management and insurance have themselves become big business.

A third and related reason is that ecological modernization explicitly avoids addressing basic social contradictions that other discourses might have introduced. Ecological modernization does not call for any structural change but is, in this respect, basically a modernist and technocratic approach to the environment that suggests that there is a techno-institutional fix for the present problems. Indeed, ecological modernization is based on many of the same institutional principles that were already discussed as solutions in the early 1970s: efficiency, technological innovation, techno-scientific management, procedural integration, and co-ordinated management.[45] It is also obvious that ecological modernization as described above does not address the systemic features of capitalism that make the system inherently wasteful and unmanageable. The leap-frog movement of capitalist innovation, periodically writing off generations of production equipment, geographical areas, and generations of workers,[46] are not addressed under ecological modernization. On the contrary, as the above suggests, ecological modernization might itself cause the writing off of generations of production equipment and industrial works.

Fourthly, ecological modernization is not merely a technical answer to the problem of environmental degradation. It can also be seen as a strategy of political accommodation of the radical environmentalist critique of the 1970s. Being the antithesis of the existing administrative judicial system, ecological modernization could mesh with the deregulatory move that typified public

[44] Schumpeter 1961: 83.
[45] For a striking illustration of this point see OECD 1985: 226, which presents a summary of the proposals for change.
[46] See Harvey 1989.

administrative thought in the early 1980s. Likewise ecological modernization had distinctive affinities with the neo-liberal ideas that were in good currency in government think-tanks and advisory agencies during the 1980s, especially concerning the need to restructure the industrial core of the economy of Western countries. As will be shown in Chapter 3, in that sense ecological modernization was an effective response to the call for alternative social arrangements as put forward by the environmental movements in the 1970s. For instance, ecological modernization straightforwardly rejects the anti-modern sentiments that were often found in the critical discourse of social movements. It is a policy strategy that is based on a fundamental belief in progress and the problem-solving capacity of modern techniques and skills of social engineering. Contrary to the radical environmental movement that put the issue on the agenda in the 1970s, environmental degradation is no longer conceptualized as an anomaly of modernity. There is a renewed belief in the possibility of mastery and control, drawing on modernist policy instruments such as expert systems and science. Furthermore, radical environmentalists, especially in the early days, were not solely committed to clean production and a restoration of respect for the integrity of nature but saw this as intertwined with a concern about increased self-determination, decentralization of decision-making, and general human growth. It seems as if, with the emergence of ecological modernization, these couplings have been marginalized.

Although it seems fair to contend that many actors would agree that the shift towards ecological modernization as described above is indeed taking place, it is also likely that they would hold quite distinct views as to what the meaning of this shift is.[47] It might be tempting to come up with a progressive reading of history as a gradual development of ideological convergence, which is the way in which the Brundtland Report approaches the issue. Is it not a great achievement that there is now an agreed vocabulary and set of institutional procedures, that both social movements, experts, and governments seem to accept as the way forward? Yet the critics of Brundtland certainly uttered a legitimate concern when they argued that ecological modernization might well be the proverbial wolf in sheep's clothing. Is ecological modernization in fact

[47] Cf. also Hajer, forthcoming.

a rhetorical ploy that tries to reconcile the irreconcilable (environment and development) only to take the wind out of the sails of 'real' environmentalists? Is ecological modernization really any more than the confident and feasible answer to what is basically another example of inefficiency and market failure or is ecological modernization just a typical OECD initiative? Yet one may also wonder whether ecological modernization does not in fact have a much more profound meaning and could be seen as the first step on a bridge that leads towards a new sort of sustainable modern society. This openness makes ecological modernization into the intriging subject it is.

The appropriate way to arrive at an answer to such questions is to conduct empirical research. That should show, first of all, to what extent the conceptual shift to ecological modernization has really occurred and whether environmental degradation has indeed become an accepted problem for governments. The second and more profound question is how the proliferation of ecomodernist story-lines affects the actual regulation and whether it leads to the institutionalization of a new type of socio-political practices. It is most likely, that something has changed, but we need a vantage-point that allows for the differentiation between one sort of institutional response and another.

This book is intended to develop insights into the way in which the emergence of the discourse of ecological modernization has affected environmental politics. A distinction of different strategies of environmental policy as made by Jänicke provides a useful first step. He distinguishes two remedial strategies and two anticipatory strategies in environmental policy-making, as illustrated in Table 1.1.[48]

This differentiation shows that the mere recognition of environmental problems still allows for quite distinct responses. Ecological modernization would mean the shift from remedial to anticipatory strategies. Concrete case-studies should show to what extent the emergence of the discourse of ecological modernization leads to a shift from remedial to anticipatory policy-making strategies, and to what extent the recognition of certain problems leads to structural change.

This differentiation remains slightly superficial. This book starts

[48] Jänicke 1988.

TABLE 1.1. *Strategies of environmental policy-making*

Type of policy-making	
Remedial	1. Repair or compensation for environmental damage due to inherently damaging products and processes of production (e.g. financial compensation for damage)
	2. Elimination of pollution through application of filters etc. on inherently damaging products and processes of production (e.g. the application of flue-gas desulphurization equipment on coal-fired power stations to fight acid rain)
Anticipatory	3. Technological modernization[a] whereby technological innovation makes processes of production and products more environmentally benign (e.g. increased efficiency in combustion)
	4. Structural change or structural ecologization whereby problem-causing processes of production are substituted by new forms of production and consumption (e.g. energy-extensive forms of organization, developing new public transport strategies to replace private transport, etc.)

[a] Jänicke in fact here uses the term 'ecological modernization'. Since it has a far more restricted meaning than ecological modernization as used in this book it is here replaced by 'technological modernization'.

from the assumption that, depending on the discursive framing of specific emblems, the ecological dilemma potentially calls into question larger institutional practices. Above we observed that ecological modernization at least partly seems to draw on the very same institutional arrangements that brought the ecological crisis about in the first place: conceptually ecological modernization relies heavily on science, technology, and expert-led processes of change. We also observed that ecological modernization is a discourse that seeks to avoid addressing basic social contradictions. However, many theorists of the environmental conflict would suggest that it is questionable whether those basic social contradictions can be kept out of the debate on contemporary environmental problems.

Regulation in this respect could well turn out to be a fundamentally discursive activity. The question is whether we can show if such problems arise and, if so, how such problems are managed.

What we need here are the conceptual tools to differentiate between different sorts of eco-modernist practices. To be able to come to such an assessment of ecological modernization I will analyse the emergence and dynamics of ecological modernization in the context of a social theory that presents a specific interpretation of the ecological crisis. For this purpose I take the work on 'risk society' and reflexive modernization as conceived by the German sociologist Ulrich Beck as point of reference.[49]

Beck's Risk Society

Beck holds that the present ecological problems signify the emergence of a new type of societal arrangement that he describes as 'risk society'.[50] Risk society, according to Beck, basically connotes a new phase in the modernization process that follows the historical era of the industrial society. In the context of industrial society, modernization referred to the renouncing of traditional arrangements and the quest for the domination of nature. The label 'risk society' is meant to indicate a forced shift in the objective of modernization from the distribution of wealth and the mastery of external threats to the management of dangers that are the inherent by-product of industrial society itself. Thus environmental crises are equated with the immense risks that are inherent in the nuclear industry, biotechnology, etc. Risk society surfaces where the emphasis in politics shifts from the distribution of goods to a—forced—concern with the distribution of bads. In risk society the flip-side of progress, the unwanted side-effects and externalities of industrial society, become a central concern.

Beck's risk society theory can be seen as a radicalization of Perrow's 'normal accidents' thesis which showed how catastrophes are an inextricable part of the complex human-technical processes of industrial society.[51] Beck suggests that the ecological deficit of

---

[49] Beck 1986 (trans. 1992), 1988, 1991, 1992; Beck *et al.* 1994.

[50] In actual fact Beck's argument is more inclusive and also comprises the transformation of the organization of social life, but since this 'individualization thesis' constitutes a separate argument with less relevance to the study of environmental politics it will not be discussed here; see Beck 1986, pt. 2.

[51] Perrow 1984.

industrial society directly backfires on the institutions that have been erected over the course of industrial modernization. The new nuclear, chemical, genetic, or ecological dangers do not respect geographical or periodic boundaries; they are hard to compensate, are often excluded from any sort of insurance, and can only partially be attributed to specific actors in terms of causality, guilt, or liability. Hence the risk society thesis would suggest that the practices on the basis of which environmental politics has so far been made, have to be fundamentally rethought. New environmental problems such as acid rain or global warming are not mere incidents that can be regulated in an incremental way. In fact they cast doubt on the social basis of the central institutions of modern society: science, the legal system, representational political institutions, and the market economy. Science is implicated because it contributed to the causes of the ecological problems and has been the prime source of legitimation (for instance in the case of air pollution, where science legitimated the so-called 'tall stack' solution to the urban smog problem of the 1950s, and thus heavily aggravated the acid rain problem). Legal arrangements have consequently individualized the attribution of guilt, which allowed for the effects to accumulate to their present proportions. Democratic institutions have been unable to guard the collective good of environment or amenity against the pressure of private gain, while the externalization of costs made the exploitation of nature into a legitimate source of economic profit. These institutional practices can now be seen to touch upon their (social) limits.

In his more recent work Beck has sought to elaborate on the risk society thesis by developing a general theory of reflexive modernization.[52] In its most simple form, the argument here is that 'the further the modernization of modern societies proceeds, the more the foundations of industrial society are dissolved, consumed, changed and threatened.'[53] This embraces many different social spheres but in the context of environmental politics one can

[52] See Beck 1993 and esp. Beck's argument in the debate in Beck *et al.* 1994.
[53] Beck in Beck *et al.* 1994: 176. For non-sociologists it is easy to overlook the theoretical significance of this argument. Beck's point is that sociological theory is essentially a product of industrial society. Hence nearly all its conceptual tools are made to analyse the internal dynamics of industrial society but these tools do not help to get the ecological problem in focus, simply because the effective domination of nature is assumed in social thought. Beck, on the other hand, suggests that the special significance of the ecological crisis for social theory is that it shows the incapacity of the central institutions of industrial society to regulate this crisis.

observe this structural sort of reflexive modernization where the modus operandi of the central institutions of industrial society that deal with environmental matters calls these institutions themselves into question. In Beck's theory this historical phase of a reflexive modernization involves both chances and threats. The ecological crisis shows the incomplete nature of the institutions of modernity. These institutions might have been successful in the provision of an unprecedented wealth, but this came at great ecological costs. Now the question is whether society can find ways of acting upon this realization of the incomplete nature of modernity. Here the ecological crisis, so Beck argues, might become the stepping-stone to a new and superior sort of modernity. Yet this is by no means the obvious outcome.

Where social structures are loosened and political legitimation becomes problematic, new societal coalitions may manifest themselves that do not see the ecological crisis as a reason for more rationality, for more democracy, and more conscious social choice. Such a 'counter-modernity' (*Gegenmoderne*) is an option that Beck takes very seriously indeed. Here Beck points out two tendencies in particular: the tendency towards new variations of ecological fundamentalism and the erosion of the political realm from within. This latter tendency is manifest in the fact that in our societies the power of the centralized institutions of parliamentary democracy has been greatly reduced precisely on those issues that are most fundamentally involved with matters of life and death, such as the risks of mega-accidents, large-scale environmental degradation, or those inherent in the employment of medical high tech or DNA research. On those issues, according to Beck, political decision-makers effectively become subordinate to expert rule.[54]

What makes Beck's analysis different from most proclamations of the techno-corporatist tendency in modern society is that he sees both tendencies as part of the emergence of what he calls

[54] Beck's explanation for the endurance of the social consensus on this techno-scientific politics of progress has three elements. Until recently the dominant perception was that technical progress equalled social progress and thus the results legitimated the regime. Furthermore there was a consequent separation of the normative debate on the social consequences of development from the technicalities themselves. And, finally, this consensus could more easily be kept up since it was actively supported by the so-called 'social partners' who formed the core of the economy and profited most from the social arrangements: see Beck 1986: 324 ff.

'subpolitics'. Subpolitics refers to the structural displacement of important political decisions to other, formally non-political, realms. Politically important decisions are in fact often taken in places that are excluded from the definition of politics one would find in classical textbooks, such as the concealed worlds of laboratories, of scientific councils (e.g. in the definition of what constitutes state-of-the-art technology or with the definition of exposure limits of certain chemicals). Subpolitics also occurs when the activities of an environmental movement or a media campaign by some actors results in a redefinition of a political issue, which then becomes the input of the political process. Beck signals a growing discrepancy between the increasingly symbolic nature of activities of traditional politics and the increasing material effects of concealed and individualized subpolitical practices. Herewith subpolitics is somewhat similar to a Foucaultian analysis of power, where one would give emphasis to the study of the combined effects of various micro powers or power/knowledge rather than to the study of the activities of a single 'sovereign' (see Chapter 2).

In this book I will analyse contemporary environmental discourse against the background of the Becksian theory of reflexive modernization. We will see environmental politics as a site where the established institutions of industrial society are put to the test. The emergence and effects of ecological modernization are analysed in this light. This means that we will try to come to grips with the specific way in which an issue which potentially threatens major social and institutional commitments is dealt with politically. For this purpose we will not only study the political process of a more traditional sort but also include the study of the subpolitical processes that Beck calls attention to.

Admittedly, the framing of this study itself seems to give part of the answer. That is to say, the emergence of ecological modernization can be seen as an attempt to take the sting out of the tail of radical environmentalism in order to secure various social and institutional commitments. Yet mere reference to a new policy discourse will not do. What needs to be analysed is how such an abstract set of concepts and ideas can come to have real political effects on the regulation of the ecological crisis. What is more, for discourse-theoretical reasons I would insist that discourses such as ecological modernization are extremely hard to control and can therefore have unintentional effects of all sorts. Hence we need to devise a conceptual apparatus that allows for the empirical analysis of the dynamics of discourses in politics, which is the task undertaken in Chapter 2.

The focus of analysis in this book will be on the practices that construct the policy problems and their solutions. The special interest is to see to what extent the practices involved are seen to have a reflexive potential, that is to say we want to see to what extent they operate in their 'industrial' routine and to what extent they allow for the creation of new conceptual combinations and lead to new institutional practices. My usage of the term 'reflexivity' is thereby substantially different from Beck's. Beck conceives of reflexivity as the self-confrontation of society or unintentional self-endangerment and distinguishes this from 'reflection' which refers to the knowledge one may have of these processes. Drawing on the insights of discourse-theory (which will be discussed in Chapter 2) I see reflexivity in the first instance as a relational notion that should be seen as a quality of discursive practices in which actors engage. Such practices are reflexive if they allow for the monitoring and assessment of the effect of certain social and cognitive systems of classification and categorization on our perception of reality. Reflexivity can thus be a quality of a metaphor or story-line that in a given context changes the perception of future perspectives. The simple metaphor of 'pollution prevention pays', for instance, illuminates the cognitive barriers of traditional business practice and makes new understandings of sound business practice thinkable. The employment of a story-line can also lead to a reinterpretation of the past, as for instance, with story-lines such as Chernobyl or Three Mile Island that allow for the reassessment of the virtues of the conceptual construct of 'residual risk', upon which the legitimate usage of nuclear power was based. But reflexivity may also be a consequence of the introduction of dissident voices in established institutional routines which interrupts the routinized way of seeing in a specific institutional realm. The reflexivity of actors is thus related to the extent to which they are able to mobilize and participate in practices that allow for the recognition of the limits to their own knowledge-base.

This outline of this research makes clear the central presupposition of this study, which is that ecological problems do not pose institutional problems by themselves, but only to the extent that they are constructed as such. Problems can be conceptualized in such a way that they pose an institutional challenge, they can be scaled down so as to become institutionally manageable incidents, or they can be seen as processes of structural change that are beyond human intervention. This is what makes the study of environmental discourse into the crucial activity it is.

2

Discourse Analysis

2.1. INTRODUCTION: THE 'COMMUNICATIVE MIRACLE'

It has become almost a platitude to characterize public problems as socially constructed. Where political analysts employ social constructivism they typically argue that public problems can in fact be related to specific 'problem owners'[1] and try to illuminate how certain actors have successfully imposed their definitions of a problem on others. In this conceptualization, social constructivism most certainly made a substantial contribution to political science. Although it is rarely related to the mainstream literature, indeed it is usually positioned in contradistinction to it, social constructivism has come up with the building blocks for a refinement of the established theory of organization as mobilization of bias. Schattschneider's celebrated definition held that 'All forms of political organisation have a bias in favour of the exploitation of some kinds of conflict and the suppression of others because organisation is the mobilisation of bias. Some issues are organized into politics while others are organized out'.[2] Social constructivism would refine this definition of the mobilization of bias, arguing that it is at least as relevant to observe the more subtle process in which some definitions of issues are organized into politics while other definitions are organized out.[3] In so doing, social constructivism substantiates Lukes's critique of the conceptions of power in the mainstream literature,[4] pointing out the lacuna left by positivist methodologies that failed to stipulate the problem of representation. They also raised new questions. How does the political play on representations take place? What does organization mean in this context, and should we be looking for 'masters of definition' akin to the gatekeepers in conventional analysis? Can we reduce the representational game to active agency as the questions

---

[1] See Gusfield 1981.   [2] Schattschneider 1960: 71.
[3] Recently this has also been argued by Torgerson 1990.
[4] Lukes 1974.

above suggest, or should we be looking for structural forms of bias as well, for instance by examining the specific idioms in which problems are discussed?

A refinement of the concept of the mobilization of bias is especially valuable in the face of the paradox of the new environmental conflict, where everybody agrees that the issue of environmental decline deserves more attention but policies do not match social expectations. The political conflict is hidden in the question of what definition is given to the problem, which aspects of social reality are included and which are left undiscussed. In this respect social constructivists have shown that various actors are likely to hold different perceptions of what the problem 'really' is. The single problem–single answer model of politics has convincingly been criticized[5] and social constructivist ideas are now widely accepted, especially due to the influence of the work of authors like Kuhn, Berger and Luckmann, Douglas, and Giddens.[6] Considering the theoretical sophistication of the work of these and other authors, social constructivist political analysis could be taken much further than the pragmatic amendment of Schattschneider's definition of the mobilization of bias suggests. In this chapter we will try to do just that, devising an 'argumentative' analytical frame for the study of political processes.

Discourse analysis has come to mean many different things in as many different places.[7] In the social sciences discourse analysis emerged in the context of the wider post-positivist interpretative tradition[8] but in fact has deep historical roots in the analysis of ideology,[9] rhetorics,[10] the sociology of science,[11] and language

[5] See Schwarz and Thompson 1990.

[6] Kuhn 1970; Berger and Luckmann 1984; Douglas 1982, 1987; Giddens 1979, 1984.

[7] Note that the often cited 'Handbook of Discourse Analysis' in fact presents only a limited section of what presently runs under the heading of discourse analysis and leaves, for instance, the most central social scientific operationalizations undiscussed: See Van Dijk 1985.

[8] Cf. Gibbons 1987 and Bernstein 1976. In essence, interpretative social science is based on a refutation of the transposition of the search for causality and the uncovering of general laws characteristic of the natural sciences, and aims instead to elucidate the meaning of certain social processes in society and to trace the various conceptual connections.

[9] McLellan points to the German tradition in Hegel and Marx that saw societies as unstable entities in which conflicting groups held conflicting perceptions of truth: see McLellan 1986.

[10] Aristotle is the obvious example.

[11] Merton 1970, 1973; esp. Fleck 1979.

philosophy.[12] Here discourse analysis primarily aims to understand why a particular understanding of the environmental problem at some point gains dominance and is seen as authoritative, while other understandings are discredited. This is taken on to analysing the ways in which certain problems are represented, differences are played out, and social coalitions on specific meanings somehow emerge.

In everyday speech, discourse is seen as synonymous with discussion, or is at best understood as a 'mode of talking'. Yet from a social scientific point of view it makes sense to reconsider this common-sense understanding of discourse. Analytically we try to make sense of the regularities and variations in what is being said (or written) and try to understand the social backgrounds and the social effects of specific modes of talking. First by analysing in which context a statement is made or to whom statements are directed. Discourse is then seen as internally related to the social practices in which it is produced. One may also point to the content of what is said. A discourse is then seen as an ensemble of ideas, concepts, and categorizations. My argument here will be that the discourses that dominate the definition of environmental problems are best analysed by combining both approaches. Discourse analysis then has a clear institutional dimension. Discourse is here defined as a specific ensemble of ideas, concepts, and categorizations that are produced, reproduced, and transformed in a particular set of practices and through which meaning is given to physical and social realities. As such, physics is an example of a discourse, but the radical environmentalists have their own discourse too. The former is produced, reproduced, and transformed through practices like academic teaching, laboratory experiments, and peer-reviewed journals. The latter is produced through the actual practising of an alternative life-style, independent protest meetings (instead of lobbying), reference to Walden or the noble savage, a specific myth about the nature of nature, and the negation of a culture of commercialism and consumption.

Coherence is not an essential feature of discourse. Some discourses, like legal discourse, will show more coherence and regularity than others, but to explain that coherence one refers to the

[12] Much recent social scientific work on discourse analysis owes a great deal to the 'linguistic turn' brought about by the later work of Wittgenstein and J. L. Austin.

(routinized) practices through which a specific discourse with its own inherent criteria of credibility is produced. Law then comes out to be a particularly well institutionalized discourse. Its high degree of coherence stems from (and is dependent upon) the reproduction and routinization of cognitive commitments in practices like the lawsuit, the legislative practice, and especially explicitly reflective practices like jurisprudence or the appeal to a higher court. In policy contexts coherence is equally dependent on its institutional environment and cannot be assumed. This makes policy discourses different from the concept of policy theory, which does assume coherence independent of context.[13]

Environmental discourse aptly illustrates the above and should not be understood as one coherent whole. In fact a discussion of a typical environmental problem involves many different discourses. First of all, phenomena like acid rain are of such complexity and are circumscribed with so much uncertainty that any single unified natural scientific discourse falls short of presenting a satisfactory understanding. At best it comes with statements that explain an element of the problem. What is more, understanding the acid rain problem not only involves the understanding of the ecological phenomenon (which in itself requires the combination of pockets of knowledge from many different disciplines) but also involves questions of cost, abatement techniques, analysis of social and economic repercussions of the different remedial strategies, and ethical questions concerning fairness or the attribution of blame and responsibility. Hence, apart from natural science discourses, many additional disciplinary stocks of knowledge—such as accounting, engineering, social sciences, and philosophy—are drawn upon. Consequently, a policy document on acid rain may easily involve discursive elements from disciplines as various as physics, tree physiology, terrestrial ecology, mathematical modelling, economics, accounting, engineering, and philosophy. One of the most striking features of the academic literature on the subject is that the repercussions of this extraordinary discursive complexity are rarely perceived to be an important topic for political scientific research. Indeed, Ackermann and Hassler are among the very

---

[13] Typical examples of policy-making practices are: the publication of White Papers, selective expert advice, informal consultation, the deliberate leaking of information to the press to monitor the social response up to the bureaucratic practice of dividing reality into smaller component parts to make it controllable.

few who consciously refer to the importance of the co-ordination of this 'bewildering variety of specialties' in environmental politics.[14]

Environmental politics brings together a great variety of actors who not only all have their own legitimate orientations and concerns, but have their own modes of talking too. The communicative miracle of environmental politics is that, despite the great variation of modes of speech, they somehow seem to understand one another. The practices in which a problem like acid rain is discussed show individual actors that contribute specific elements of knowledge and/or considerations. These ideas may make perfect sense within the discourse in which they were constituted, but then subsequently become an element in a debate that is conducted by a far more diverse group of actors and in the context of acts and practices that do not function according to the discursive logic of that original discourse. Acid rain is an example of an inter-discursive issue. The question that thus emerges is how the different actors involved find ways to communicate at this inter-discursive level and how the many separate elements of knowledge are related to come to form authoritative narratives on acid rain. What is more, we want to know how social power is exercised in this context. The second part of the agenda is to investigate how permanence is secured and change comes about. What can discourse theory teach us about the shifting conceptualizations of environmental problems?[15]

In this chapter I first seek to advance a theoretical understanding of discursive dynamics by discussing two different discourse-theoretical approaches. First I will discuss the work of Michel Foucault and subsequently I will deal with the potential contribution of the social psychologists Billig and Harré. Subsequently I define the discourse-theoretical components that can underpin the analysis of the dynamics of discourse-formation in the two acid rain controversies.

---

[14] Ackermann and Hassler 1981: 4.

[15] It should be observed that this communicative miracle characterizes more than environmental politics alone. The increasing specialization and rationalization have created more and more specific modes of talking. At the same time all this specialist discourse is seen as an essential input into 'proper' decision-making. Hence the problems of knowledge transfer described above can be seen as an essential feature of a highly modern society.

2.2. THE INTERACTION OF DISCOURSES

There is an important divide that distinguishes the discourse theory that underpinned Foucault's early work from his later studies.[16] It is his later work that is especially usefully to this study. The two main studies of his later work analysed the social discourses on social discipline and punishment, and sexuality. Foucault broke these discourses down into the multiplicity of component discourses that were produced through a whole array of practices in various institutional contexts. One of the essential arguments here was that proper analytical research should focus precisely on the illumination of the smaller, often less conspicuous practices, techniques, and mechanisms, which he called 'the disciplines'; these somehow determined how large institutional systems actually worked. So, for instance, in *Discipline and Punish* he sought to show how the legal system that was devised in the spirit of the philosophy of the Enlightenment 'guaranteed a system of rights that were egalitarian in principle [but] was supported by these tiny, everyday, physical mechanisms, by all those systems of micro-power that are essentially non-egalitarian and asymmetrical.'[17] Foucault criticized political theory for paying too much attention to institutions and too little to these smaller practices or 'disciplines'.[18]

The focus on the plurality of discourses led Foucault to criticize the historical practice which sought to understand history in terms of causality. Alternatively, and not necessarily less ambitiously, Foucault wanted to explain the play of dependencies between discourses, and 'to render apparent the polymorphous interweaving of correlations.'[19] Given the emphasis on the plurality of discourses, Foucault problematized the play of discontinuities within and between these discourses. He came to speak of 'different types of transformation' rather than of a univocal process of social change, and emphasized the need to investigate the 'micro-powers' that

---

[16] For this point see especially Dreyfus and Rabinow 1986. In this section we are not concerned with the historical content or political theory in Foucault's work but with his methodology of discourse analysis. The most useful guides here are Foucault's lectures on governmentality (Burchell *et al.* 1991; Foucault 1968), his *Discipline and Punish* (1975), Volume One of his *History of Sexuality* (1976), and perhaps the best discussion of his work, the aforementioned Dreyfus and Rabinow (1986).

[17] Foucault 1975: 222.   [18] See Gordon 1991: 4.

[19] Foucault 1975: 58.

brought about transformations. Nevertheless, he was very insistent that these transformations happened according to definable
rules, that there was a 'discursive order' that could be illuminated.
This discursive order would illuminate the regulated discursive
practices through which objects are constituted as communicable
entities in a given society.

The role of the subject in this discursive order has often been a
cause of puzzlement. Foucault was keen to falsify the idea that the
relationship between the individual and discourse should not be
conceptualized as that of a 'sovereign subject' that manipulates
passive discursive structures. One may note that such radical
voluntarism has been widely criticized in relation to institutional
practice, but somehow language is often strangely overlooked.
The common sense of the sociology of knowledge these days is
that (institutional) structures are both constraining and enabling.[20]
Research should therefore focus on the interaction between agency
and structure and the way in which interaction transforms the
rules of domination.[21] Foucault argued that

> there are not on the one hand inert discourses . . . and on the other hand,
> an all-powerful subject which manipulates them, overturns them, renews
> them; but that discoursing subjects form a part of the discursive field—
> they have their place within it . . . and their function . . . Discourse is not
> a place into which the subjectivity irrupts; it is a space of differentiated
> subject-positions and subject-functions.[22]

Foucault emphasized that discourses are not a mere set of signs
but contain internal rules that make discourses function as a structure to behaviour in the sense of what Giddens called the duality
of structure: as enabling and constraining.[23] A problem with
Foucault's analysis is that his discourse theory often seems intermingled with his historical argument. Foucault's emphasis on discipline as the dominant theme of modernization is parallelled by

---

[20] See Berger and Luckmann 1984; Giddens 1979, 1984.

[21] See Giddens 1984: xx. Giddens, being inspired by Foucault, obviously does
not exclude discourse from his theory of structuration.

[22] Foucault 1968: 58.

[23] See also Foucault 1976: 101. As we will see below, this makes discourse
analysis far more sophisticated than traditional research on ideology. Discourse
analysis allows for the study of bias that might have ideological effects but might
not have had ideological intentions.

a heavy emphasis on the constraining workings of discourse, but is rather weak on the enabling aspect. His understanding of the constraining effects of discourse is nevertheless instructive. Discourses imply prohibitions since they make it impossible to raise certain questions or argue certain cases; they imply exclusionary sytems because they only authorize certain people to participate in a discourse; they come with discursive forms of internal discipline through which a discursive order is maintained; and finally there are also certain rules regarding the conditions under which a discourse can be drawn upon.[24]

The role of the subject was seen as conditional upon the discursive field (see the quote above) in which various positions and functions of the subject were inscribed. Power was not a feature of an institution (i.e. the Sovereign) but was defined relationally, referring to the way in which institutions and actors are implicated in discourses (in which inequalities, disequilibria, divisions, and other categories are defined).[25] Hence in this view the power of an institution is permanent in so far as it is a constant feature of the discourses through which the role of that institution is being reproduced.[26]

In this context Foucault called for the study of the 'positivity' of discourses, that is, the way in which discourses are conditional upon certain rules regarding the conditions under which they emerge, exist, and how discourses transform. It was Foucault's alternative to an approach that disregards discourse and judges 'what is said by who says it, or vice versa.'[27] Hence, according to Foucault there is no a priori thinking subject trying to express or transcribe his or her preconceived ideas in language. The subject operates in the context of a whole group of regulated practices according to which his or her own ideas are formed. Historical

[24] See Foucault 1971.

[25] For instance, the fact that people in the Western world (tend to) stop in front of a policeman or a traffic-light is not an illustration of the power of that individual policeman—let alone the traffic-light—but of the discourse (in which the relevant rules and conventions are contained) in which the role of the policeman or the traffic-light are defined.

[26] Foucault 1976: 93.

[27] Foucault 1968: 69. Foucault's historical work on scientific discourses thus basically became a demystification of discourses like science, showing how they too were dependent on certain political practices and circumstances: see especially Foucault 1966.

research should illuminate how individual actors act within and upon these practices.[28]

Finally, consider Foucault's idea of the 'tactical polyvalence of discourses.' It refers to the way in which the various discursive elements, that might have been introduced for various unrelated strategic purposes, together create a new discursive space within which problems could be discussed. Here the structuring capacity of discourse gains its meaning within the context of consciously operating actors. In *The History of Sexuality* Foucault suggested that we must imagine the world as

a multiplicity of discursive elements that can come into play in various strategies. It is the distribution that we must reconstruct, with the things said and those concealed, the enunciations required and those forbidden, that it comprises; with the variants and different effects—according to who is speaking, his position of power, the institutional context in which he happens to be situated—that it implies; and the shifts and reutilisations of identical formulas for contrary objectives that it also includes.[29]

Individual strategic action, thus far played down or conspicuously absent, is brought back in and relates to the level of discursive practices that Foucault endeavoured to understand. Foucault is very insistent that agency should be considered in the context of discourses that enable and constrain action:

the rationality of power is characterised by tactics that are often quite explicit at the restricted level where they are inscribed (the local cynicism of power), tactics which, becoming connected to one another, attracting and propagating one another, but finding their base of support and their condition elsewhere, end by forming comprehensive systems: the logic is perfectly clear, the aims decipherable, and yet it is often the case that no

---

[28] According to Foucault's later work discourse is more than 'a surplus which *goes without saying*, [that] does nothing else except say what is said.' (Foucault 1968: 63). Even more appropriate might have been to end the statement with 'except say what has already been conceived'.

[29] Foucault 1976: 100, emphasis added. An example of reutilization can be found in Foucault's account of the disqualification of homosexuality in the 19th cent. This took place via the invention of a whole set of new discursive concepts in psychiatry, jurisprudence, and literature that constructed homosexuality as 'perverse', and situated it apart from society. Yet what Foucault notes is that this negative discourse was subsequently reversed: the same principles were later used by homosexuals to emphasize their own identity and to call for the legitimation of homosexuality. It illustrates that discursive struggle is not a matter of two separate discourses competing: it is about the way in which specific discursive elements are related.

one is there to have invented them, and few who can be said to have formulated them.[30]

In conclusion, the strength of Foucault's theory of discourse lies at the level of discursive practices and the interaction and coalescence of discourses, but the role of the discoursing subject remains ambivalent. One should take into consideration, however, the specific questions Foucault sought to answer: his research was concerned with the developments of social discourses during the modern era, which obviously reduces the importance of interpersonal discursive (inter)action. On the other hand the focus on micropowers was an often reiterated methodological principle and in that light it is somewhat unsatisfactory not to analyse the power-effects at the restricted, tactical, interpersonal level but only at the aggregate level that is suggested in the above quotation.

Foucault shares the view of post-positivist political science that suggests that discourse is not to be seen as a medium through which individuals can manipulate the world as conventional social science suggests. It is itself part of reality, and constitutes the discoursing subject.[31] Foucault's theory of discourse shows that the reference to institutional backgrounds or vested interests is an unsatisfactory circular explanation because institutions are only powerful in so far as they are constituted as authorities *vis-à-vis* other actors through discourse. Similarly, interests cannot be taken as given a priori but are constituted through discourse. The point here is that interests have to be constantly reproduced and will change over time, for instance through what Foucault described as the play of discontinuities between discourses. The task of the political analyst will be to explain how a given actor (whether it is an organization or a person) secures the reproduction of his discursive position (or manages to alter this) in the context of a controversy. The influence of a stubbornly resisting actor, then, cannot be explained by reference to the importance of his position alone, but has to be given in terms of the rules inherent in the discursive practices, since they constitute the legitimacy of his position.

Foucault's theoretical concepts are a valuable source of inspiration

---

[30] Foucault 1976: 95.

[31] This constitutive view of language is widely endorsed: for an overview see Bernstein 1976.

for the study of discourse formation in politics. Yet there is a need to devise middle-range concepts through which this interaction between discourses can be related to the role of individual strategic action in a non-reductionist way. Likewise, there is still a conceptual gap between Foucault's abstract work and the study of concrete political events. Furthermore Foucault lacks a proper theory of permanence and change. In the present operationalization a notion like the polyvalence of discourses unduly emphasizes coincidental recombinations, but one wonders whether there cannot be a middle ground between the *epistèmes* of Foucault's early work that mysteriously governed processes of discourse formation, and the contingency of the power effects that is evident in his later work.

2.3. THE IMPORTANCE OF STORY-LINES

How can we operationalize the effect of discourse on the workings of various practices in a given domain? How can individual discourses, which are operational in distinct quarters of a domain, come to influence one another? What role is there for the subject in the context of a notion like Foucault's polymorphous interweaving of correlations? Here the discourse-theoretical ideas that have recently been developed in the field of social psychology can help to define a discourse-analytical approach that is both theoretically sophisticated and practically operationable.

Authors like Harré or Billig defend a discourse-theoretical position that, to a certain extent, has an overlap with the work of Foucault.[32] Yet they take part in a debate that focuses on a completely different level of abstraction, namely the level of interpersonal interaction. I will call this perspective 'social-interactive' discourse theory. It potentially contains some valuable corrections of Foucault's theory. Like Foucault, social-interactive theory uses a constitutive view of language.[33] Here this is used to question the dominant views within the social-psychological discipline. Human

[32] See Harré 1993; Davies and Harré 1990; Billig 1989, Billig *et al.* 1988, but see also Potter and Wetherell 1987; Edwards and Potter 1992.

[33] It should be observed that this idea that language constitutes reality always refers to the Lacanian level of reality and necessarily excludes the level of the Real; see Sect. 1.2.

interaction is not related to roles and ritualized social practices, it is argued, but to discursive practices in which people are provided with what they call 'subject-positions'. The critique of the role concept is that it assumes that a person is always separable from the role taken up. Yet actors can only make sense of the world by drawing on the terms of the discourses available to them. Alternatively, the social interactionists argue that persons are constituted by discursive practices, and they conceptualize human interaction as an exchange of arguments, of contradictory suggestions of how one is to make sense of reality. Research should examine the specific discursive practices defined as 'all the ways in which people actively produce social and psychological realities.'[34]

Billig, who works in the spirit of Harré, goes back to the ancient theorists of rhetorics and shows how they already dealt with these kind of questions.[35] The object of research is the practices through which actors seek to persuade others to see reality in the light of the orator or rhetorician. These authors thus argue not so much for a linguistic turn (examining discursive systems) but for an argumentative turn:

> To understand the meaning of a sentence or whole discourse in an argumentative context, one should not examine merely the words within that discourse or the images in the speaker's mind at the moment of utterance. One should also consider the positions which are being criticized, or against which a justification is being mounted. Without knowing these counter-positions, the argumentative meaning will be lost.[36]

If we examine controversies in environmental politics from this perspective these conflicts are not to be conceptualized as semi-static plays in which actors have fixed and well memorized roles of environmentalist, policy-maker, scientist, or industrialist. On the contrary, environmental politics becomes an argumentative struggle in which actors not only try to make others see the problems according to their views but also seek to position other actors in a specific way. Hence it is not as if actors do not have an intuitive idea about discourse theory, in actual fact they constantly practice it.

Here we can infer the first major correction to Foucaultian

---

[34] Davies and Harré 1990: 45.
[35] For similar attempts see also Engbersen *et al.* 1991.
[36] Billig 1989: 91.

discourse theory. The argumentative interaction is a key moment in discourse formation that needs to be studied to be able to explain the prevalence of certain discursive constructions. The social-interactive perspective sees actors as 'active, selecting and adapting thoughts, mutating and creating them, in the continued struggle for argumentative victory against rival thinkers.'[37] From this perspective, Harré and Billig argue for the analysis of 'witcraft', the skills of argumentation. Based on their work one could aim to discern the crucial claims in a particular issue, what is seen as a persuasively structured argument, what style of presentation is effective, or what historical positioning serves to justify a particular course of action. In a similar vein, Engbersen points to the Aristotelian distinction between *logos*, *ethos*, and *pathos*. This could help to explain the political influence of certain arguments or claims. *Logos* refers to how to argue a persuasive case, *ethos* refers to the reputation of the speaker, and *pathos* is a rhetorical strategy that aims to play on the emotion of listeners.[38] Engbersen argues that in present-day politics one can distinguish specific legitimate argumentative formats that govern a particular domain. Billig's work could likewise inform the analysis of what one may call 'discursive styles' that are characteristic of modern bureaucratic arguing. For instance, his rhetorical approach sensitizes research to the effects of 'categorization' according to which a particular issue is processed as just another element of a general category, or of 'particularization', where the uniqueness of a case is emphasized.

Discourse analysis then investigates the boundaries between the clean and the dirty, the moral and the efficient, or how a particular framing of the discussion makes certain elements appear as fixed or appropriate while other elements appear problematic. One can endeavour to show whether definitions 'homogenize' a problem, that is to say make the problem understandable within a reified perception of the wider problem field, or whether definitions suggest a 'heterogenization' that requires an opening up of established discursive categories.[39] To deconstruct a policy discourse and find that it is to be understood as the unintended consequence of an interplay of actions is one thing, more interesting

---

[37] Billig 1989: 82.   [38] Engbersen 1991: 145.
[39] Cf. Billig's terms 'categorisation' and 'particularisation', above.

is to observe how seemingly technical positions conceal normative commitments, yet more interesting still is to find out which categories exactly fulfilled this role, and which institutional arrangements allowed them to fulfill that role, i.e. how this effect could occur and which course of affairs is furthered in this way.

So far this exposition of the social-interactive discourse theory has presented ways in which the subject can be studied as actively involved in the production and transformation of discourse. It thus fills a gap left by Foucault. At the same time both perspectives are compatible, in principle, because they both work from a relational ontology and focus on the study of 'practices'.

The second corrective to Foucault refers to the perspective on social change and permanence. One of the theoretical assumptions of the 'positioning' theory of Davies and Harré is that discourse is reproduced through a sequence of speech situations. The fact that there are similarities between statements (i.e. historical continuity) is to be explained by memory or historical references that people draw upon in a new 'speech situation'. This so called 'immanentist' view has interesting implications. Marx's often quoted dictum that people make their own history but not under conditions of their own choosing is mostly interpreted to mean that actors are not totally free to act since they have to cope with the existing social structures. The immanentist view of language of Davies and Harré shows that this political context is also to be analysed as a discursive construction. Rules, distinctions, or legitimate modes of expression, only have meaning to the extent that they are taken up. It implies that the rules and conventions that constitute the social order have to be constantly reproduced and reconfirmed in actual speech situations, whether in documents or debates. Consequently, the power structures of society can and should be studied directly through discourse.

This has interesting consequences for the research of politics and policy-making. Analysing interpersonal communication thus becomes much more relevant. Analysing policy papers becomes important even if they do not include 'hard' new proposals or legislation. It becomes imperative to examine the specific idea of reality or of the status quo as something that is upheld by key actors through discourse. Likewise it becomes essential to look at the specific way in which appositional forces seek to challenge these constructs. Discourse analysis, then, is not only essential for the analysis of subject positions but also for 'structure positionings' (referring to which structural elements can be changed, and what institutions remain to be seen as fixed or permanent).

Change and permanence thus come to depend on active discursive reproduction or transformation. In this process actors are not totally free but are, as holders of specific positions, entangled in webs of meaning. Here the routinization of cognitive commitments gives permanence to discursive understanding: 'Once having taken up a particular position as one's own, a person inevitably sees the world from the vantage point of that position and in terms of the particular images, metaphors, story lines and concepts.'[40]

Social-interactive discourse theory combines the appreciation of the importance of routinized understandings with the appreciation of the possibility on the part of specific actors to exercise—at least a notional—choice in relation to the various practices available to them. In this context Davies and Harré introduce the concept of story-lines (albeit without properly defining it), which to my mind hints at a subtle mechanism of creating and maintaining discursive order. A story-line, as I interpret it, is a generative sort of narrative that allows actors to draw upon various discursive categories to give meaning to specific physical or social phenomena. The key function of story-lines is that they suggest unity in the bewildering variety of separate discursive component parts of a problem like acid rain. The underlying assumption is that people do not draw on comprehensive discursive systems for their cognition, rather these are evoked through story-lines. As such story-lines play a key role in the positioning of subjects and structures. Political change may therefore well take place through the emergence of new story-lines that re-order understandings. Finding the appropriate story-line becomes an important form of agency.[41]

Hence, although this social-interactive theory of discourse has a firm idea of a subject, it does not argue that all action and positioning is the result of an active process of taking up or denying of positionings. On the contrary, there is a considerable power in structured ways of seeing. These are often based on reification, that is to say, their arbitrary character remains hidden. Consequently, people do not recognize them as moments of positioning

[40] Davies and Harré 1990: 46.
[41] We will come back to story-lines in Sect. 2.4, below.

but simply assume that this is '"the way one talks" on *this sort of occasion.'*[42] Routinized forms of discourse thus express the continuous power relationship that is particularly effective because it avoids confrontation.[43] The relevance of this discussion to environmental politics will be obvious. The environmental movement is haunted by the dilemma of whether to argue on the terms set by the government or to insist on their own mode of expression. In the latter case, of course, they run the risk of loosing their direct influence and therefore they often barter their expressive freedom for influence on concrete policy-making. The social-interactive perspective raises the question to what extent actors recognize the way in which the government talks as a reification. Davies and Harré emphasize that the subject can always deny the terms set by the initial speaker and emphasize the importance of the availability of alternative discourses. The disciplinary force of discursive practices often consists in the implicit assumption that subsequent speakers will answer within the same discursive frame. Even if they do try to challenge the dominant story-line, people are expected to position their contribution in terms of known categories. Discursive challenges may consist of withstanding understandings in terms of routinized categories or, often even more powerful, in establishing new combinations within seemingly traditional discursive structures (e.g. by introducing new historical examples). This would be an example of how the discoursing subject can actively exploit the tactical polyvalence of discourse.

It should be observed, however, that Davies and Harré give relatively little attention to the degree to which discourse can become structured in institutional arrangements.[44] In political reality, to argue against routinized understandings is to argue against the institutions that function on the basis of specific, structured, cognitive commitments. This does not make the discursive interaction look like a debate between clever, creative individuals, but frustrates debate in the rather dumb role-playing on the part of those whose action takes place within walls of routinized institutional structures.[45] Although it seems true to say that institutions

---

[42] Davies and Harré 1990: 49.   [43] See also Beetham 1991: 45.

[44] This is partly due to the social psychological context within which the concept of positioning was introduced and the particular agenda of its intervention; see the beginning of this section.

[45] See Lloyd 1986: 272.

function only to the extent that they are constantly reproduced in actual practices, these routinized institutional practices tend to have a high degree of salience.[46]

A perhaps more serious limitation to the usefulness of the positioning perspective for political science can be derived from Habermas's argument that the intersubjective element is precisely what has been marginalized in capitalism and is increasingly being replaced by money power.[47] Focusing on the intersubjective moment might therefore obscure the understanding of the real power relationships. However, even money power assumes some sort of discursive interchange, whether as a threat to withdraw investments or, more likely, as anticipated reaction, in a discussion among dependent actors. In the former case the relevant discursive practice is the exchange between a firm and the state, in the latter case the relevant discursive practice is localized within government (in cabinet meetings, parliamentary debates, or in departmental deliberations). Both can be investigated.

2.4. THE ANALYSIS OF DISCOURSE-COALITIONS

From the two traditions presented above I derive the social-theoretical foundation of a new approach to the analysis of discourse in political contexts. It directs attention primarily to the socio-cognitive processes in which so called 'discourse-coalitions' are formed.[48] My 'argumentative' approach[49] focuses on the constitutive role of discourse in political processes as described above and allocates a central role to the discoursing subjects, although in the context of the idea of duality of structure: social action originates in human agency of clever, creative human beings but in a context of social structures of various sorts that both enable and constrain their agency. The transformational model of social reality then maintains that society is reproduced in this process of interaction between agents and structures that constantly adjusts, transforms, resists, or reinvents social arrangements.[50]

---

[46] See Douglas 1987.   [47] Habermas 1986: 390 ff.

[48] The notion 'discourse-coalition' is also developed and used by Peter Wagner and his collaborators (Wagner 1990). Although his conceptualization is likewise derived from a reflection on Giddens, Bourdieu, and Foucault there are some important differences which I will indicate below.

[49] See Forester and Fischer, 1993; Billig 1989.   [50] See Giddens 1984.

In so doing, I aim to position myself first of all against studies that see social constructs as a function of the interests of a group of actors. In that case language is seen as a means and it is assumed that actors use language purely as a passive set of tools. In actual fact there is much more interaction between the linguistic structures and the formation of preferences. In this book language is seen as an integral part of reality, as a specific communicative practice which influences the perception of interests and preference. Interests, as my account of Foucault and the social psychologists has shown, cannot be assumed as given. Interests are intersubjectively constituted through discourse. This has important repercussions for the study of environmental politics, since it suggests that the emergence of a new policy discourse like ecological modernization may actually alter the individual perception of problems and possibilities and thus create space for the formation of new, unexpected political coalitions.

Secondly, and related to the argument above, my approach should be distinguished from theories that ground their argument in the idea that actions and perceptions should be understood against the background of deeply held beliefs or belief systems. Here each belief system has its own a priori way of seeing and its own way of arguing things. The argumentative approach focuses on the level of the discursive interaction and argues that discursive interaction (i.e. language in use) can create new meanings and new identities, i.e. it may alter cognitive patterns and create new cognitions and new positionings. Hence discourse fulfils a key role in processes of political change.

The argumentative approach conceives of politics as a struggle for discursive hegemony in which actors try to secure support for their definition of reality. The dynamics of this argumentative game is determined by three factors: credibility, acceptability, and trust. Credibility is required to make actors believe in the subject-positioning that a given discourse implies for them and to live by the structure positionings it implies; acceptability requires that position to appear attractive or necessary; trust refers to the fact that doubt might be suppressed and inherent uncertainties might be taken for granted if actors manage to secure confidence either in the author (whether that is an institute or a person), e.g. by referring to its impeccable record, or in the practice through which a given definition of reality was achieved, e.g. by showing what sort of deliberations were the basis of a given claim. Discursive dominance or hegemony is thus seen as an essentially socio-cognitive product, whereby the social and the cognitive are seen as essentially intertwined. Hence arguments can convince because of some property they have—e.g. plausibility—that countervailing ideas lack, but one has to reckon that in such cases plausibility is the product of persuasion which is not a purely cognitive process.

This struggle does not take place in a social vacuum but in the context of existing institutional practices. One should analyse in which practices discursive dominance is based and by what means specific contentions are furthered. In other words, institutional arrangements are seen as the pre-conditions of the process of discourse-formation. Yet the organization of science, of policy-making, or of democratic procedures does not determine this process: from Foucault and Harré we derive the insight that institutions (and the practices that constitute them) need discursive 'software' to operate and produce effects. In terms of research this implies that we should find out how institutions are made to operate through subject positionings and structure positionings that lend closure to an institutional machinery that can be put to different uses.[51] In this respect discourse analysis should illuminate two things. First, the way in which cognitive and social commitments are routinely reproduced. Second, the way in which discursive 'interpellations' take place, whereby interpellations are understood as those moments where routinized proceedings are interrupted. The focus of analysis here is on those discursive practices which, drawing on the work of Harré, can be understood as the inter-discursive transfer points where actors exchange positional statements and where new discursive relationships and positionings are created.

Discourse has been defined as a specific ensemble of ideas, concepts, and categorizations that is produced, reproduced, and transformed in a particular set of practices and through which meaning is given to physical and social realities. We will speak of the condition of discourse structuration if the credibility of actors in a given domain requires them to draw on the ideas, concepts, and categories of a given discourse, for instance, if actors' credibility

---

[51] Obviously, some institutional arrangements can also be found to have a structural bias against certain applications.

depends on the usage of the terms of ecological modernization in the domain of environmental politics. We will speak of discourse institutionalization if a given discourse is translated into institutional arrangements, i.e. if the theoretical concepts of ecological modernization are translated into concrete policies (i.e. shifting investment in mobility from road to rail) and institutional arrangements (introduction of multi-value auditing, or the restructuring of old departmental divisions).[52] If these two conditions are satisfied, a discourse can be said to be hegemonic in a given domain.[53]

In the introduction to this chapter we observed that the debates on issues like acid rain are essentially inter-discursive in nature. This was meant to underline the fact that an understanding of the phenomena necessarily requires the combination of knowledge claims that are the product of distinct discourses. This has a profound influence on the sort of communication that typically takes place in the context of environmental issues. To be able to analyse this inter-discursive communication the argumentative approach puts forward the concepts story-line and discourse-coalition as middle-range concepts that can show how discursive orders are maintained or transformed. An essential assumption in a discourse-coalition approach is that the political power of a text is not derived from its consistency (although that may enhance its credibility) but comes from its multi-interpretability. Consider once more the communicative miracle that accompanies the discussion of inter-discursive problems like acid rain. Although many actors make their own contribution to the understanding of the acid rain problem in its full complexity, there are hardly any actors (if any at all) who can actually understand the problem in all its details. Characteristic of the discourse on acid rain is the fact that knowledge becomes politically relevant once it is transcribed into a higher order (political) discourse. There is thus a constant need to generate ways of reproducing e.g. scientific findings in non-scientific discourse. Here Schön has pointed to the widespread usage of generative metaphors in politics.[54] Schön argued that metaphors

---

[52] Note the difference with Wagner, who refers to discourse structuration akin to Giddens's theory of structuration as the process of reproduction and transformation of discourses in society (Wagner 1990: 33, 55).

[53] Obviously, in practical politics this is usually only the case to a certain degree. Such a weaker form of hegemony might be labelled 'discursive domination'.

[54] Schön, 1979.

provide a common ground between various discourses.[55] Actors are thus given the opportunity to create their own understanding of the problem, re-interpreting various elements of knowledge outside their specific realm of competence, or filling in the gaps and ambivalences that were left by the original text. This is the interpretive process of 'discursive closure' in the course of which complex research work is often reduced to a visual representation or a catchy one-liner. Extreme cases are a single graph representing the longitudinal development of forest damage or the reduction of ten years of research in air chemistry in a discursive change from 'tall stacks will dilute and disperse emissions' to a saying like 'what goes up must come down'. It is obvious that this translation is accompanied by a loss of meaning: in both cases all uncertainty and all conditionality of the original knowledge claims is erased to come to discursive closure. Argumentative discourse theory goes one step further and argues that regulation depends on this loss of meaning and the multi-interpretability of text.

Drawing on Davies and Harré I argue that the regulation of conflict over inter-discursive problems like acid rain depends on and is determined by the effects of certain story-lines. Story-lines are narratives on social reality through which elements from many different domains are combined and that provide actors with a set of symbolic references that suggest a common understanding.[56] Story-lines are essential political devices that allow the overcoming of fragmentation and the achievement of discursive closure. As Popper already acknowledged, fragmentation is a more general characteristic of investigatory practice (thus effectively discrediting the prevalent metaphor of filling the gaps of knowledge).[57] The point of the story-line approach is that by uttering a specific element one effectively reinvokes the story-line as a whole. It thus

---

[55] See also Lakoff and Johnson 1980.

[56] The idea of examining the structuring role of story-lines in policy contexts is also used by Rein and Schön, who speak of 'problem setting stories', often based on generative metaphors, that participants tell about policy situations, and that relate specific causal accounts to certain proposals for action: Rein and Schön 1986. However, they base these story-lines on the belief systems that people share. This makes their approach vulnerable to the critique that can be formulated on the belief system approach (see the discussion on Sabatier, below).

[57] 'With each step forward, with each problem we solve, we not only discover new and unsolved problems, but also discover that where we believed that we were standing on firm and safe ground, all things are, in truth, insecure and in a state of flux.': Popper quoted in Billig 1989: 72.

essentially works as a metaphor.[58] First of all story-lines have the functional role of facilitating the reduction of the discursive complexity of a problem and creating possibilities for problem closure. Secondly, as they are accepted and more and more actors start to use the story-line, they get a ritual character and give a certain permanence to the debate. They become 'tropes' or figures of speech that rationalize a specific approach to what seems to be a coherent problem. Thirdly, story-lines allow different actors to expand their own understanding and discursive competence of the phenomenon beyond their own discourse of expertise or experience. In other words, a story-line provides the narrative that allows the scientist, environmentalist, politician, or whoever, to illustrate where his or her work fits into the jigsaw.

Argumentative discourse analysis holds that the power of story-lines is essentially based on the idea that it sounds right. This should not be misunderstood as a purely cognitive process. Whether something sounds right is not only influenced by the plausibility of the argument itself, but also by the trust that people have in the author that utters the argument and the practice in which it is produced and is also influenced by the acceptability of a story-line for their own discursive identity.

The concept of story-line is a key element in the argumentative approach. Story-lines fulfil an essential role in the clustering of knowledge, the positioning of actors, and, ultimately, in the creation of coalitions amongst the actors of a given domain. Story-line is the analytical term that unites several established concerns in research in the constructivist tradition. The discursive practice of the metaphor, recently rediscovered in political science, for instance, comes under the definition of a story-line, as do analogies, historical references, clichés, appeals to collective fears or senses of guilt. These shallow and ambiguous discursive practices are the essential discursive cement that creates communicative networks among actors with different or at best overlapping perceptions and understandings. They are, therefore, also the prime vehicles of change.

---

[58] Notice the difference with Wagner's theory of discourse. He does not use the concept of story-line but suggests that a similar process of 'implicit connotations' takes place in the context of discourses (where using one element carries with it implicit connotations for the discourse as whole) (Wagner 1990: 24). In other words, he suggests a much more stable cognitive basis than my suggestion that coalition formation is based on largely associative and metaphorical understandings.

Consider the case of acid rain. In essence 'acid rain' is a narrative that relates certain industrial emissions to the dying of fish, lakes, and trees and the corrosion of buildings. If phrased in purely scientific idiom, acid precipitation will most likely only have a limited effect. However, as a causal story that gives meaning to previously singular and unrelated events such as dead fish, dying trees, and smoking stacks, it might have a significant power-effect. What is more, it changes the meaning of elementary phenomena like smoking stacks or dirt. As O'Riordan observed, smoke and dirt had, until recently, a radically different meaning. They were interpreted as signs of progress and wealth, as is indicated by the Yorkshire expression 'Where there's muck, there's brass.'[59] The acid rain story-line thus alters what is seen as 'out of place' and creates new insights into the social order. Consequently, the story-line might give rise to new political claims. If it were to be made available to him, the acid-rain narrative could provide the local fisherman who experienced the demise of his catches and had seen this as a matter of fate, with the conceptual apparatus to perceive the reality of dead fish in an alternative way. Similarly, it would confront the forester with a plurality of possible realities.[60] If the forester noticed needle loss or discoloration of leaves before, he or she would have had to see it as the product of natural stress caused by drought, cold, or wind. With the acid-rain story-line the forester might also see it as a result of pollution. What is more, once he or she has become familiar with the acid rain narrative the forester that had so far not been aware of widespread damage in his district, might change his or her way of seeing reality, in this case the forest: occurrences that he or she had previously conceived as evolutionary, might be interpreted as evidence of pollution. The discourse of pollution is thus empowering in the sense that it gives the fisherman and forester a focus for protest and the argumentative ammunition to argue their case. Story-lines, in other words, not only help to construct a problem, they also play an important role in the creation of a social and moral order in a given domain. Story-lines are devices through which actors are

---

[59] O'Riordan 1985. A well-known Dutch expression has it that *De schoorsteen moet roken* ('The stack should smoke') and Ashby and Anderson (1981) likewise point to the fact that in the Victorian Era it was thought that smoke spelt employment. See also Ashby and Anderson 1976: 289–90.
[60] See Gusfield 1981.

positioned, and through which specific ideas of 'blame' and 'responsibility', and of 'urgency' and 'responsible behaviour' are attributed. Through story-lines actors can be positioned as victims of pollution, as problem solvers, as perpetrators, as top scientists, or as scaremongers.

The second middle-range concept of the argumentative approach is discourse-coalitions. The argumentative approach holds that in the struggle for discursive hegemony, coalitions are formed among actors (that might perceive their position and interest according to widely different discourses) that, for various reasons (!) are attracted to a specific (set of) story-lines. Discourse-coalitions are defined as the ensemble of (1) a set of story-lines; (2) the actors who utter these story-lines; and (3) the practices in which this discursive activity is based. Story-lines are here seen as the discursive cement that keeps a discourse-coalition together. The reproduction of a discursive order is then found in the routinization of the cognitive commitments that are implicit in these story-lines (cf. Davies and Harré, above). Discourse-coalitions are formed if previously independent practices are being actively related to one another, if a common discourse is created in which several practices get a meaning in a common political project.

Ecological modernization can be used to illustrate the concept. Ecological modernization is based on some credible and attractive story-lines: the regulation of the environmental problem appears as a positive-sum game; pollution is a matter of inefficiency; nature has a balance that should be respected; anticipation is better than cure; and sustainable development is the alternative to the previous path of defiling growth. Each story-line replaces complex disciplinary debates. What is more, story-lines imply arbitrary confinements: they often conclude debates that are still open or imply a marked move away from an academic consensus.[61] However, certain actors are attracted to the idea of ecological modernization that, as a discourse, presents a new set of ideas, concepts, and categorizations through which they can give meaning to the physical and social realities that are implied in the contemporary environmental conflict. The influence of the new policy discourse depends on the cognitive power of its story-lines, but also on its

---

[61] An example of the latter case is the way in which ecology is drawn upon in policy-making. As Kwa has pointed out, in academic circles the idea of a 'balance of nature' was refuted in the 1960s: see Kwa 1987.

attractiveness. As I showed in Section 1.2, the environmental issue potentially raises rather substantial questions about the social order. Yet here ecological modernization suggests that the environmental issue can be remedied without having to completely redirect the course of social developments.

Discourse-coalitions here differ from traditional political coalitions or alliances. First of all, in its emphasis on the linguistic basis of the coalition: story-lines, not interests, form the basis of the coalition, whereby story-lines potentially change the previous understanding of what the actors' interests are. Secondly, it broadens the scope of where the participating actors are to be located. Discourse-coalitions suggest searching for politics in new locations, looking for the activity of the actors who produce story-lines (i.e. scientists, activists, but also mediators such as journalists), and the practices within which this takes place, for instance by investigating the role of a popular scientific magazine in the construction or proliferation of a story-line, or by looking at the activities of specific organizations in bringing together previously independently operating academics or policy-makers.

Thus far we have argued that story-lines play an essential role in reproducing and transforming a discursive order in a given policy domain. However, the fact that story-lines reduce the discursive complexity does not explain why actors from various backgrounds adhere to them. Empirical research will have to illuminate the specific strategic reasons why actors introduce or support specific metaphorical understandings. But that alone would present too individualist an explanation. It is important to recall some of the more structural notions from Foucault that were introduced earlier, most notably the 'tactical polyvalence of discourses' and 'polymorphous interweaving of correlations'. If we consider the workings of story-lines from that perspective we can see how they essentially operate on the middle ground between *epistèmes* and individual construction. We would argue that story-lines do not primarily derive their discursive power from the individual strategic choice, or from the fact that the specific elements fit together in a logical way, but because they depend on what I propose to call 'discursive affinities.'[62] Separate elements might

[62] The idea of affinities comes from Weber's notion of 'elective affinities' that refer to practices that mutually favour the continuance of the other practice (Weber, 1968).

have a similar cognitive or discursive structure which suggests that they belong together. In that case actors may not understand the detail of the argument but will typically argue that 'it sounds right'.[63] This element of the explanation of a discursive order thus does not primarily refer to the actors and their intention but explicitly operationalizes the influence of discursive formats on the construction of problems.

If a discursive affinity is particularly strong I propose to call this 'discursive contamination'. In that case discursive elements not only resemble one another but flow over into one another. A famous example of discursive contamination in environmental discourse is the case of Darwinism. Darwin drew on sociological concepts such as 'competition' and used them as a metaphor to understand the natural reality. The discursive contamination later gained an extra dimension as the Social Darwinists subsequently drew on Darwinian thought to argue that competition was the natural state of society.[64]

Story-lines not only provide the narrative within which a specific actor can understand his specific contribution to knowledge or localize his or her own social preference in the context of specific scientific findings. Story-lines can also be seen to influence actors in their own production of knowledge. An important example here is the way in which the 'precautionary principle', one of the principle story-lines that structured the discourse of ecological modernization, influenced the discourse of biological science. Gray has shown how the precautionary principle started to appear in scientific reports.[65] The precautionary principle is of course not a natural scientific concept. It is a policy principle which was introduced as the antithesis of its predecessor: remedial environmental politics. It was meant to illuminate the credibility of the idea of anticipatory policy and to create new coalitions in pollution politics. In that context the precautionary principle holds that policymakers will sometimes have to decide on action even if there is no scientific evidence of a causal link (the so called 'no regrets scenario'). Gray, however, demonstrates how scientists introduced the precautionary principle in the way in which they reported on

---

[63] Whetherell and Potter come to similar conclusions (1988: 170). Yet note the difference with Wagner's concept of 'cognitive affinity', which suggests a much stronger relationship.
[64] See Worster 1991.   [65] Gray 1990.

their findings. They started to draw different conclusions from their data than could have been drawn on the basis of the accepted statistical norms and hence altered the scientific practice and contaminated scientific discourse. It shows that the unification of different fields is not a pragmatic alliance that leaves reality unchanged. On the contrary, in the process of creating new discursive understandings practical reality might change too.

2.5. CONCLUSION: A COMPARISON WITH ADVOCACY-COALITIONS

Above I have presented the outlines of my argumentative approach to discourse. Where, then, does the argumentative approach and the analysis of discourse-coalitions differ from more conventional approaches to the study of policy-making as a process of coalition-formation? To illustrate this I will discuss the differences with Sabatier's conceptual framework of 'advocacy-coalitions'.[66] In so doing I compare discourse-coalitions with what I perceive to be an extremely challenging approach. At least some of the disagreements with Sabatier can be overcome and this book hopes to provide a contribution to this debate.

Sabatier argues that policy change is best analysed as a struggle between competing 'advocacy-coalitions' at the level of a policy subsystem (i.e. all organizations that are concerned with a given problem). Advocacy-coalitions are made up of people from various organizations 'who share a set of normative and causal beliefs and who often act in concert.'[67] The policy-making process is seen as a struggle between coalitions that is eventually mediated through the activities of 'policy brokers'. This results in certain decisions taken by a 'sovereign'. These decisions produce effects that, in turn, result in social learning concerning the perceived influence of external dynamics, or the nature of the problem at hand.[68] This cognitive learning in the policy subsystem is not the only source of change. There is also change in the real world both outside the subsystem (e.g. changing socio-economic parameters, changing coalitions in government) and inside the subsystem (e.g. changing personnel) that help to bring about policy change.[69]

---

[66] Sabatier 1987.   [67] Ibid. 652.
[68] Cf. the figure ibid.   [69] Ibid. 654.

I agree with Sabatier on several major points. It makes sense to focus the investigation on the process of coalition formation at the level of a policy subsystem or policy domain. Recent work in political science and sociology suggests that the play of political forces within a given policy domain (defined as those organizations that deal with a specific substantive issue) is the most important determinant of the way in which substantial policy issues are dealt with.[70] Structuring ideas are generated and put to work in the close interaction that takes place between various actors within that domain. I also support Sabatier's decision to choose to analyse the political interplay in such domains through the analysis of distinct coalitions, but differ on the nature of these coalitions. Finally, I would agree that the controversies between these coalitions should always be understood against the background of external parameters, including relatively stable parameters like constitutional structures, social structures, or geographical predispositions, or, indeed, the economic dependence of governments.

There are three major points were Sabatier's advocacy-coalitions can be seen to be essentially different from discourse-coalitions: (1) the individualist ontology differs from my relational ontology; (2) the central role of beliefs in advocacy-coalitions differs from my emphasis on the constitutive role of language and the role of story-lines and discursive affinities; (3) Sabatier's notion of policy-oriented learning differs from my theory of social change. There is, however, a major overlap between the two approaches, but this can at least partly be explained by the fact that Sabatier's ideas seems to touch upon the limits of his methodology. Let me explain.

First, Sabatier holds an individualist ontology. An advocacy-coalition is made up of certain scientists, policy-makers, politicians, and journalists who share a belief-system. Sabatier presents advocacy-coalitions as an analytical alternative to the analysis of institutions and actors. The latter option is refuted because the large number of organizations active in a subsystem would make this task too complex.[71] Yet the individualist ontology is hardly an improvement. The analysis of practices suggests that what people say differs according to the practice in which they engage. Their values and beliefs might differ accordingly. A person might be a

---

[70] Burstein 1991.   [71] Sabatier 1987: 660–1.

perfect father in the context of his family practices but a tough businessman in the board-room, or even a serial killer after he has put the children to bed. We are likely to find similar differences in a more moderate variation if we analyse the way in which individuals behave in the context of policy-making. Looking at the beliefs of individuals as variables to explain policy change fails to consider the importance for the process of change of the varied nature of the contributions made by specific individuals in different argumentative exchanges. For instance, the fact that actors utter contradictory statements implies that their activity may help sustain different coalitions. Taking over the new story-lines of a rival coalition (even if this is only to criticize that position) acknowledges the existence of the alternative perspective and may thereby facilitate the reproduction of that coalition.

The problems with the individualist ontology also come out in the boundary problem of his framework. An advocacy-coalition is defined as individuals who share a belief system. Later he admits that some individuals in a coalition do not share the belief system: they may participate 'simply because they have certain skills to offer, but otherwise may be indifferent to the policy disputes.'[72] It is hard (if not impossible) to draw a boundary here: who shares the beliefs that define the coalition and who is only professionally involved?[73] The boundary problem also comes out in the distinction between 'advocates' and 'brokers'. Policy-brokers mediate between coalitions and search for acceptable solutions. Sabatier uses high-level civil servants in Britain as example to explain this category.[74] He subsequently argues that many brokers (including high-level civil servants) may have some interest in policy, while advocates may sometimes be concerned with conflict accommodation: this is defined to be an empirical question. Most likely empirical research on what actors actually say will show that some elements should be seen as advocacy while other statements are attempts to be a policy broker. Here Sabatier seems the victim of his own individualist methodology. He can only resolve the contradiction by opening up his ontological stand: actors 'play the role of policy broker.'[75] The question then is whether the play in

---

[72] Sabatier 1987: 662. Incidently, at one point Sabatier himself identifies a 'foundation' as a broker: see ibid. 678.

[73] It also falsely assumes that technical skills do not have political implications.

[74] Sabatier 1987: 678.   [75] Ibid. 662.

which the role is performed stretches over the ten years of the controversy, i.e. whether actors always play the same parts. It seems more plausible that actors that operate within a given coalition position themselves or are being positioned as brokers at certain moments in the course of a specific argumentative interplay while being advocates at other moments or places.

Second, the cement of advocacy-coalitions is the shared beliefs of individuals. Sabatier takes these beliefs as a priori, although they are changeable through social learning. In both cases language is seen as a means. In the first case language is drawn upon to express the stable values of an individual, in the second case rational reflection on policy developments may change personal insights. In the argumentative approach both language and context help to constitute beliefs. People are not seen as holding stable values but as having vague, contradictory, and unstable 'value positions'. New discourses may alter existing cognitive commitments and thus influence the values and beliefs of actors, for instance because new story-lines create new cognitions that may give people a new idea about their potential role and the possibilities for change (i.e. new subject and structure positionings).

Third, Sabatier's notion of policy-oriented learning illuminates a strong rationalist idea about cognitive change. Sabatier identifies as a key question how people can learn across belief systems. He distinguishes three important factors.[76] First, learning is facilitated by informed conflict. This implies that both sides must have the appropriate facilities. He also argues that debates must try to avoid discussing core questions to avoid defensive reactions that hinder reasoned debate. Secondly, learning requires a relatively apolitical forum 'in which the experts of the respective coalitions are forced to confront each other.'[77] Central variables for success then are prestige, professional norms, and peer review. Thirdly, learning is easier with problems for which quantitative performance indicators can be found and if natural systems are concerned because the critical variables are not behaving strategically and controlled experimentation is more feasible.[78] This reveals a fundamentally different idea of what social conflicts over problems like air pollution are about and how these social conflicts can be

---

[76] Ibid. 678–81.   [77] Ibid. 679.
[78] The ideas on policy-oriented learning are central to Sabatier's approach and constitute four of the nine hypotheses he draws up in his article.

resolved. Sabatier's approach rests on the assumption that scientific debate can bring about a consensus on policy matters on the basis of an exchange and comparison of objective findings. However, studies in the sociology of science in policy contexts have shown how scientific claims are often intermingled with policy claims.[79] Other studies have argued that experts arguing from different premises cannot come to consensus (they argue from 'contradictory certainties') but need external influence to come to social closure.[80] The type of instruments of social learning that Sabatier suggests have been shown to be instruments of regulation but in quite a different way than Sabatier suggests. In his study into the Windscale Inquiry Wynne has revealed how the demands of rationalism and objective science facilitate the formulation of some beliefs and values while defining others as irrelevant.[81]

The argumentative approach operationalizes the idea that discourse is constitutive of the realities of environmental politics. The environmental conflict thus does not appear as primarily a conflict over which sorts of action should be taken (or whether action should be taken) but as a conflict over the meaning of physical and social phenomena. In this process story-lines fulfil a key role. They determine the interplay between physical and social realities. Story-lines are seen as the vehicles of change and are analysed in connection to the specific discursive practices in which they are produced. This methodology will help to explain the political dynamics in the environmental domain.

[79] Cf. Wynne, 1987; Jasanoff 1990.
[80] Ravetz 1990; Schwarz and Thompson 1990.
[81] Wynne 1979, see also Ch. 6.

3

The Historical Roots of Ecological Modernization

3.1. INTRODUCTION

To be able to understand the present political dynamics of ecological modernization we will have to go back to at least the early 1970s when the environment quite suddenly became a political topic in Western societies. Since that day environmental discourse has taken many twists and turns, from the collective concern about the prophecies of doom of a coming resource crisis that would bring the whole world to a grinding halt, to the localized concern over pollution of water, soil, and air; from the concerns about the possibility that the modern world will come to an end with the bang of a nuclear catastrophe to the idea that Western civilization is slowly ruining its heritage now cathedrals and sculptures crumble away as the consequence of acid rains. The historical sequence of emblematic issues is often retold in the literature. Yet there is a parallel story that is not often told. While the alleged catastrophes made the headlines and have always dominated both the academic and popular literature, there was another process that laid the foundations of the present age of ecological modernization. Here I want to reconstruct the specific argumentative interplay between the state, the environmental movement, and key expert organizations that made ecological modernization into such a powerful force. Hence I have no interest in trying to summarize all the complex currents of environmental politics and ecological thinking over the last twenty-five years.[1] Likewise, it should be obvious that this argumentative interplay should be understood against the background of changing socio-economic parameters, such as the changing political and economic climate, and the confrontation with the unprecedented scaling up and speeding up in

[1] For this, cf. O'Riordan 1983; Dobson 1990; Eckersley 1992; Goodin 1992.

industrial (and particularly chemical) production after World War Two. What we will focus on here is the institutional process out of which the new discourse of ecological modernization could derive its social support. This institutional genealogy will help to explain the specific emphasis and orientations that now characterize ecological modernization and the dynamics of the debates on the regulation of emblematic issues like acid rain in the 1980s or global environmental problems in the 1990s.

This institutional approach should correct the prevailing image of environmental politics, according to which the widespread political awareness of the troublesome state of the environment is the result of the recognition of the true extent of environmental degradation. Politically powerful reports like *Our Common Future* suggest that there has been a growing awareness of the cumbersome state of affairs that has culminated in the new consensus on the need for action and has thus provided the basis for the emergence of ecological modernization. Many studies have shown this one-dimensional view to be wrong. Most recently Von Prittwitz introduced the helpful notion of a 'paradox of catastrophes' to make this point.[2] Using the example of urban air pollution, Von Prittwitz shows how the concern over pollution grew as pollution diminished. Then, drawing on the case of radiation protection, he shows how awareness and political activities were constantly less significant at places where radioactive contamination was most intense. Clearly, environmental politics is more dynamic than the linear account of the Brundtland Report would have it. Yet, as we will show, this is not the line that is taken in much of the literature. What often fails is the analysis of how certain macro-changes in value-orientations affect patterns of institutionalization or indeed concrete policy processes. It is that intermediate level that we will address in this chapter.

3.2. THE ORIGINS OF ENVIRONMENTALISM

Present-day environmentalism has its roots in the late 1960s. Environmentalism, which is here described as the social concern over environmental change, emerged as an element of a much

---

[2] See Von Prittwitz 1990.

broader value change that occurred at that time. Probably the best known research on the changing value-orientations in the late 1960s is Ronald Ingelhart's work on the 'silent revolution.'[3] Inglehart observed a marked shift in value preferences from goals that related to material values (concerning economic and physical security) towards goals that related to so called 'post-material' values (concerning human growth and the satisfaction of intellectual and aesthetic needs). He argued this on the basis of large-scale survey research in Western countries. It is relevant to examine this theory in some detail, since Inglehart's work has often been used to explain the emergence of environmentalism in the late 1960s.[4] To understand his argument one must recognize three assumptions of his research.[5] First, Inglehart draws on Maslow's theory of the hierarchy of needs.[6] This implies that people are assumed to fulfil their material needs first and only then to aspire to post-material goals. Secondly, his research presupposes that people value most highly those things that are in short supply: the principle of diminishing marginal utility. Thirdly, he argues that the more or less definitive personal ordering of values takes place during the formative years of childhood and youth. The 'silent revolution' holds that the post-war generation, which experienced peace, economic stability, and affluence during their youth, held a totally new set of 'post-material' values. Their parents, on the other hand, having experienced both the depression of the 1930s and the Second World War, were likely to hold more materialist values. This, then, is Inglehart's explanation for the emergence of what Roszak called the 'counter-culture' of the 1960s and Galtung calls the Green movement.[7]

Where Inglehart's work is drawn upon for the explanation of the emerging environmentalism in the late 1960s and 1970s it is characteristically argued that the post-material generation constituted the core support for the environmental movement.[8] However, even though Inglehart has refined his thesis over time, his influential work can be criticized on many grounds.[9] As far as his

---

[3] See Inglehart 1971, 1977, 1984, 1990.

[4] See Lowe and Goyder 1983; Cotgrove 1982; Cotgrove and Duff 1980; O'Riordan 1983.

[5] See Lowe and Rüdig 1986: 515.   [6] Maslow 1954.

[7] Roszak 1969; Galtung 1986.   [8] Cotgrove 1982; Eckersley 1989.

[9] See Lowe and Rüdig 1986: 517.

assumptions are concerned, it has been pointed out that Maslow admits that there are problems with his own hierarchy of needs which obviously also affects Inglehart's work. The fulfilment of basic needs does not automatically lead to higher order need fulfilment.[10] Furthermore, the emphasis on childhood socialization has been questioned. According to Eckersley, education, and especially the higher educational experience, is a more important factor contributing to environmentalism. It has also been noticed that the environmental movement in fact had a rather heterogeneous make-up comprising many different groups.[11] Yet the main reason why Inglehart's thesis does not suffice as a contextual basis for this study is that it gives an idealist explanation that focuses entirely on 'push factors', as Offe calls them, rather than on 'pull factors'.[12] Inglehart's theory completely separates the psychological development of post-material values from actual social, economic, and indeed environmental developments. As Lowe and Rüdig have pointed out, that divorce makes the Inglehart thesis the wrong starting-point for an explanation of environmentalism and environmental politics.

An attempt to escape Inglehart's idealism could elaborate on the scarcity hypothesis in his work. Here Fred Hirsch's seminal work on the social limits to growth could function as starting-point.[13] Hirsch emphasized that the good things in life are restricted not only by physical limits but also because the perception of the quality of many of these goods diminishes as more people have access to them. Hirsch calls this social scarcity. For instance, if hiking through the desolate valleys of Scotland is recognized as an aspect of the good life and more and more people take up walking, the quality of the individual experience will go down. This is also true for goods such as higher education and urban amenities. The satisfaction of consumption of such 'positional goods' is dependent on their non-possession by others.[14]

This social scarcity hypothesis, then, could be drawn upon to provide a more structural foundation for the explanation of the

[10] See Eckersley 1989: 216; for a critique of Maslow also see Thompson *et al.* 1990: 55 ff.

[11] Including ones that do not belong to the post-material value cohort: see Cotgrove and Duff 1980; Cotgrove 1982; Offe 1985.

[12] Offe 1985; see also Rohrschneider 1988.

[13] Hirsch 1977.   [14] See Crouch 1983: 186.

increase in environmentalism.[15] During the 1960s a well-educated new middle class emerged, that held post-material values and aspired to positional goods. Following this line of reasoning, it was the experience of social scarcity that made this group increasingly aware of the negative effects of growth and critical of the idea that economic growth equals social progress. The merit of Hirsch's argument is that he relates the changing values to their structural consequences. This can subsequently be used to provide an explanation for the increased reflexive awareness of the environmental issue as an manifestation of these social limits to growth. However, a potential problem with an application of Hirsch's work to the emergence of environmental values is that his work is based on the assumption that individuals are selfish and anthropocentric. This does not match the character of much of the environmentalist concern of that period.[16] For instance, the radical core of the environmentalists of the 1960s and 1970s, argued for wilderness for its own sake, which cannot immediately be described as fulfilling a post-material need.[17] As Hays pointed out, there is a big difference between wilderness as a recreational good and wilderness as a value in itself.[18] Hence, although Hirsch's work clearly contributes to our understanding of the environmental conflict, it is not a sufficient explanation.[19]

In general, there are several problems with the explanations of the emergence of environmentalism in the late 1960s and early 1970s that have been discussed so far. First of all, they all seem to fail to account for the great variety of expressions of environmental concern that emerged during that period. In the early 1970s it was not only the counter-culture or the new middle class that raised their voice against environmental decline. It was also the period of the UN conference at Stockholm and the publication of the influential report *Limits to Growth* by the Club of Rome.

[15] See Watts and Wandesforde-Smith 1981, and also Lowe and Goyder 1983; Eckersley 1989.

[16] See Offe 1985.   [17] See Eckersley 1989: 220.   [18] Hays 1979: 143.

[19] Hirsch's work can also serve to illustrate the élitist bias in much work on environmentalism. After all, many of the new car-owners who in the 1960s could drive out into the countryside for the first time might not have shared Hirsch's experience of congestion but would have felt it as a real improvement in their working-class lives, as Gershuny observed: see Gershuny 1983. It is another illustration of Downs's argument that 'The elite's environmental deterioration is often the common man's improved standard of living.' Downs in this respect speaks of 'the democratisation of privilege.': see Downs 1972: 44.

Although it is obviously true that the social movements helped to bring about institutional change,[20] it would be wrong to reduce the great variety of expressions of social concern to responses to the radical environmental movement. It is more likely that they were one—albeit important—element of a larger political dynamic. Secondly, the explanations discussed above are all rather static and are unable to explain the changing nature of environmentalism since the late 1960s. For that purpose we will need a different sort of analysis.

3.3. EARLY POSITIONINGS: *LIMITS TO GROWTH* AND *BLUEPRINT FOR SURVIVAL*

As one studies the literature on environmental politics one is struck by the tendency to make sense of environmental conflict through the introduction of specific taxonomies of environmental ideologies.[21] Analysts are looking for internally consistent paradigms or deeply held beliefs that motivate the political actions of specific actors. As such it reflects the influence of the belief system approach that we discussed in Chapter 2. In this search for consistency many of the contradictions and ambiguities of environmental discourse are lost. Distinctions are made more clear-cut than they are in reality, or one might be tempted to create internally consistent social paradigms where in reality contradictory notions are more likely to be found. Moreover, there is a fair chance that the taxonomy will guide the presentation of research and lead to the pigeonholing of actors while these individual actors themselves might hold, or to be more precise, utter contradictory ideas depending on the discursive practices in which they find themselves. The reconstruction of paradigms or belief systems excludes the intersubjective element in the creation of discourse. It overlooks that in concrete political situations actors often make certain utterances to position themselves *vis-à-vis* other actors in that specific situation, emphasize certain elements and play down others,

---

[20] See Offe 1985.

[21] See e.g. O'Riordan 1983; Pepper 1986; Cotgrove 1982; Schwarz and Thompson 1990; Porritt and Winner 1988: 29.

or avoid certain topics and agree on others. Depending on their discursive strategy and the process of mutual discursive positioning, actors might sometimes seek to suggest communalities to create a discourse-coalition, or emphasize differences to build up counter-discourses. Although this will be more obvious on the confined level of a specific policy process, mechanisms akin to this inter-subjective positioning through discourse also seem to have affected the general environmental debate between 1972 and the emergence of ecological modernization in the mid-1980s.[22]

Having given up the idea that one should aim to reconstruct clear ideological or policy paradigms in the environmental conflict, we have opened up the way to bringing back into the analysis all those ambiguities and indeed commonalities that a research guided by taxonomies seeks to avoid or will tend to smooth over. Another consequence is that I do not seek to present an overview of the whole political spectrum of views that could be found in environmental discourse. My orientation towards the link between discourse and regulation makes it interesting to see precisely which ideas sought to occupy the centre of the debates. By so doing we might be able to illuminate the discursive affinities and discursive positionings that were instrumental to the political emergence and sustaining of the policy discourse of ecological modernization.

A first element of an understanding can be derived from a description of the studies that dominated the centre of the debate (both having far more radical and less contradictory rivals on their other side) in the early 1970s, *Limits to Growth* on the one hand and *Blueprint for Survival* and *Small is Beautiful* on the other.[23] These three texts shared much but had a very distinct outlook at the same time. Together they created a widespread credibility for the claim that the environmental crisis was serious and needed to be addressed.[24] Yet in terms of regulation a hierarchical and technocratic top-down approach typified by the report *Limits to Growth*

---

[22] It should be observed that O'Riordan is aware of this problem. He emphasized, for instance, the centrality of the interplay between a soft-technologist's perspective and an accommodator's perspective. (O'Riordan 1983: 378).

[23] Meadows *et al.* 1972; the *Ecologist* 1972; Schumacher 1973.

[24] In the Western world, one should add, since Third World countries were highly suspicious of the sudden attention given to the environmental issue and thought this only served to push the issue of development and redistribution of wealth aside. This was one of the key conflicts at the 1972 Stockholm conference.

competed with a bottom-up participatory approach that can be found in reports like *Blueprint for Survival* and *Small is Beautiful.*[25]

The World Model of the Club of Rome

Limits to Growth examined the consequences of five major and mutually interrelated trends: 'accelerating industrialization, rapid population growth, widespread malnutrition, depletion of non-renewable resources, and a deteriorating environment.'[26] It suggested that continued exponential growth would cause a resource crisis within a period of 100 years and projected a grim neo-Malthusian vision of a rapidly growing population, food shortages, disappearing resource reserves, and increasing levels of pollution. It combined this prophecy of doom with a call for radical reform of the existing decision-making processes to achieve 'the state of global equilibrium.'[27] *Limits to Growth* was fiercely criticized not only for its methodological flaws and untenable Malthusian assumptions[28] but also for its implicit political agenda.[29] How could it have come to play such a prominent role?

To be sure, the apocalyptic message was not new. It was the ethos of the report and the general ideological climate of the early 1970s that gave the message political resonance and made the environment into a political priority. The ethos of *Limits to Growth* stemmed partly from the fact that it was published by the Club of Rome, a club of more or less prominent figures from business, policy-making, and science. Hence this time its apocalyptic predictions could not be so easily discredited as yet another product of

[25] Here we will not give an in-depth analysis of the actual content and coherence of the two reports. For a critical appraisal of *Limits to Growth* and *Blueprint for Survival*, see O'Riordan 1983: 60 ff.; Paehlke 1989: 41 ff.; the editorial in *Nature*, 235: 63; and esp. Cole *et al*. 1974. They argue that the assumptions of the MIT research may be unduly pessimistic but still warn for complacency.

[26] Meadows *et al*. 1972: 21.   [27] Meadows *et al*. 1972: 24.

[28] Critics showed that the premisses of exponential growth in all sectors made the whole computing exercise programmed to doom. See especially Cole *et al*. 1974. For the implicit Malthusian assumptions, see Sandbach, 1978: 496–501. It should be added, however, that both the MIT research group and the Club of Rome were well aware of the many uncertainties that were implicated in the World Model. For the Club of Rome this is obvious in its commentary that is included in the actual report.

[29] See. e.g. Golub and Townsend 1977 and Harvey 1974 on both aspects.

marginal, anti-progress activists.[30] The Club of Rome was a re-markable initiative. *Limits to Growth* was the first product of its Project on the Predicament of Mankind. The task which the Club of Rome had set itself was nothing less than to bring about a process of global cognitive change. As such it demonstrated clear affinity with previous attempts to try to create a shift in thinking about the power-deficit in international politics: the existing power structures seemed incapable of meeting the global challenges facing the industrial world. This issue had been raised most explicitly by the Committee for Atlantic Economic Cooperation in the late 1960s and could rejoice in widespread attention from powerful actors from the world of finance, trade, and industry.[31]

Yet the ethos of *Limits to Growth* (which was after all subtitled 'A report *for* the club of Rome') was further reinforced by some of the characteristics of its authors. It had been written by estab-lished experts from the Massachusetts Institute of Technology (MIT) who were seen as leading scientists in the promising field of cyber-netics. Here a group of experts around Jay Forrester was working on the design of a dynamic computer model that would cover all aspects of reality.[32] Under the leadership of Dennis Meadows this 'World Model' was drawn upon to provide answers to the specific questions that concerned the Club of Rome. The ethos of *Limits to Growth* was enhanced by the fact that cybernetic systems theory was very much in vogue in circles of decision-makers at that time. It was seen as a promising opportunity to extend the possibilities of a rational and scientifically based form of decision-making at a time when the response to the growing complexity of social relations became an increasing prevalent theme in governmental circles.

This combination of an apocalyptic message, wrapped up in the cybernetic discourse that was well respected, and uttered by actors who could not immediately be discredited, helps to explain why the message of environmental doom that was presented in *Limits to Growth* could have had such an unprecedented impact on élite

[30] The Club of Rome had been initiated in 1968 by Aurelio Peccei, director of Olivetti, and was funded by various industries. Among its leading members were Alexander King, then scientific director at the OECD, Eduard Pestel from the Technical University of Hannover, and Umberto Colombo, who was also attached to the OECD.

[31] The founder of the Club of Rome, Aurelio Peccei, had a great personal in-volvement in this: see Golub and Townsend 1977.

[32] See Forrester 1969, 1971.

opinions.[33] In terms of regulation the cybernetic language made *Limits* into an ideal reference for White Papers at a time at when discourse was seen by many as a productive way of thinking about complex issues. In all, *Limits to Growth* was an example of an extremely successful use of discourse as power. Despite its completely unofficial status, the aura of respectability and knowledge ability that accompanied the Club of Rome report became a key reference in the debate on state of the environment. But more important still, the resonance of *Limits to Growth* meant that others came to conceptualize the environmental problem according to a specific set of concepts and categories.

This was, undoubtedly, precisely what the Club of Rome had intended.[34] Of course, *Limits to Growth* was also meant to confront society with the possible effects of a continued reliance on the existing 'physical, economic and social relationships' in the world: the Club of Rome incessantly emphasized that it intended the report to be a catalyst for change. Yet their prophecy of doom also served to influence the direction in which the prescribed 'radical reform of institutions and political processes at all levels'[35] had to be sought. In this respect it might, in the first instance, seem surprising that the emphasis of the report was on the outline of the problems related to continued exponential growth; and that, although it argued that only a 'steady state' society would be able to prevent a global collapse, the short 'Commentary' by the Club of Rome that accompanied *Limits to Growth* left implicit how the task of changing society into a peaceful 'steady state of economic and ecological equilibrium' was to be taken on.[36] It hinted at a preferred direction, arguing that it was well aware that 'the strategy

[33] In all it sold more than two million copies. Perhaps most extremely, in the Netherlands the translated report sold a hundred thousand copies within a month.

[34] The Foreword mentioned as the purpose of the Club of Rome (aptly described as an 'invisible college') 'to foster understanding of the varied but interdependent components—economic, political, natural, and social—that make up the global system in which we all live; to bring that new understanding to the attention of policy-makers and the public worldwide; and in this way to promote new policy initiatives and action.' (Meadows *et al.* 1972: 9), italics added. Off course this intention cannot explain the success of the Club of Rome in imposing a new discursive frame on the discussion.

[35] Meadows *et al.* 1972: 193.

[36] The report itself put a heavy emphasis on the importance of technological improvements (including recycling, better product design, solar energy, medical advances, and contraceptives).

for dealing with the two key issues of development and environment must be conceived as a joint one',[37] and suggesting that the physical limits to growth 'are further reduced by political, social, and institutional constraints, by inequitable distribution of population and resources, and by our inability to manage very large intricate systems.'[38] Yet as a discursive power-practice *Limits to Growth* was perhaps effective precisely because it concentrated on the definition of the problem. After all, if you define a problem you already partly define its solutions, while you do not immediately run the risk of being accused of using a problem to further your own goals.

Limits to Growth for the first time portrayed the environmental issue as a global crisis. This construction of the problem as a world-threatening collapse already guided the search for solutions. The cybernetic model made the effort of maintaining the ecological and the social order a matter of the increased application of techniques of scientific management. These implicit recommendations of the Club of Rome became rather more explicit in later (and less influential!) work by the Club of Rome, as O'Riordan observed.[39] A 1975 publication by Mesarovic and Pestel,[40] endorsed by the Club of Rome, appeals 'for global political reconciliation, a minimum adequacy of subsistence for all mankind, a world resource management plan, and cooperative action by the multinationals to spur industrialisation and agricultural development in the third world.'[41] The study further comes out in favour of liberalization of trade and resource distribution, rejects the idea of a steady state arguing for 'organic growth' instead, and explicitly rejects the idea of self-sufficiency.[42]

In conclusion, *Limits to Growth* became the catalyst of a coalition that started from a recognition of the serious nature of the environmental conflict but which sought to remedy the environmental

---

[37] Meadows *et al.* 1972: 192.
[38] Meadows *et al.* 1972: 186.   [39] See O'Riordan 1983: 63.
[40] Pestel was himself member of the Executive Committee of the Club of Rome.
[41] O'Riordan 1983: 64.
[42] Radical critics also read *Limits to Growth* as an attempt by representatives of an international élite to help create a climate in which further redistribution of wealth and power would be blocked and new forms of global control would become acceptable, whilst restricting the independence of national governments and facilitating the aspirations of multinational firms to enhance their role in world politics: see Pepper 1986.

predicament through a further integration of organized management. The environment was in danger of becoming a runaway issue, both socially and physically, but careful planning, drawing on the latest scientific insights, could restore the equilibria. *Limits to Growth* was oriented towards the world leaders and national élites which it hoped to unite for a joint approach to the problem. Despite the serious academic critique that immediately followed its publication, *Limits to Growth* was remarkably successful in this respect.

Blueprint for Self-Government?

Blueprint for Survival came up with rather different ideas concerning the regulation of the environmental crisis. Interestingly, *Blueprint for Survival* in fact derived its urgency from the same MIT extrapolations of exponential growth and also called for a steady-state society. Like *Limits* it argued for the conversion of the flow economy into a stock economy minimizing the usage of non-renewables. Among its proposals were also many technological fixes, such as the introduction of recycling schemes, energy conservation, an ecotax, a reorientation from private to public transport, new environmentally sound agricultural practices and a new way of calculating GNP, many of which are now seen as core elements of ecological modernization.

Yet in other respects it was very much unlike *Limits*. *Blueprint* gave a complete outline of an environmentalist strategy which included much more than strictly environmental matters.[43] *Blueprint* problematized the mode of production, the existing capital–labour relations, and the lack of morality in industrialized and urbanized society. Yet the most significant difference surely was that *Blueprint* proposed to decentralize industrial society into smaller, highly self-sufficient and communal units. Most importantly, it argued for the development of 'relatively "closed" economic communities' working with intermediate technologies.[44] In many respects these proposals for a new social order, made up of

---

[43] Nevertheless, it received widespread support also from scientists. In an appeal accompaning its publication 34 prominent scientists endorsed its principles and a further 180 supported the report in broad terms: see the letter to the editor by a group of distinguished Oxford biologists in *Nature*, 235: 405.

[44] The *Ecologist* 1972: 30; for intermediate technologies, see below.

neighbourhood councils and workers co-operatives, reflected an-
archist ideas of authors like Peter Kropotkin. Here the difference
with *Limits* could hardly have been greater. *Limits* argued for a
further integration and hierarchization in order to contain the problem (global problems requiring global solutions) while *Blueprint*
argued for decentralization, self-sufficiency, and self-government;
Limits accentuated the historical responsibility of social élites of
business, science, and government, *Blueprint* came out in favour
of self-government. In all, *Blueprint* was in fact a rather contradictory romanticist critique of modern society that argued for an
anti-technocratic, decentralized utopia but at the same time drew
on cybernetics to illustrate the urgency of its call and relied on
comprehensive planning techniques to bring about the new utopia
of self-sufficiency.[45]

Historically *Blueprint* and *Limits* fulfilled similar roles for distinct audiences. *Blueprint* quickly came to be seen as the more
radical corrective to *Limits to Growth*. And whereas *Limits to
Growth* typified very much the response to the increased importance of environmental matters from the world of business and
governmental élites, *Blueprint* became a key reference for the radical environmental public of its time. It called for the formation of
a Movement for Survival of grassroots organizations and here it
quickly became linked to the classic contribution by Fritz Schumacher, *Small is Beautiful*. Schumacher criticized Western materialism, questioned the Western definition of progress, and came out
strongly in favour of an economics of permanence, a revaluation
of humanist values in Christianity and non-Western traditions such
as Buddhism and other Eastern religions: 'In the excitement over
the unfolding of his scientific and technical powers, modern man
has built a system of production that ravishes nature and a type
of society that mutilates man.'[46] Schumacher tried to describe a
feasible alternative in which a new sense of moral community was
reconciled with the best of the technological advancements. He
argued for small-scale working units based on intermediate technology that were designed on the basis of an understanding of the
laws of ecology. Schumacher's machines would serve people rather
than the other way round, and would not require a comprehensive

---

[45] It included a detailed blueprint of how the transformation was to take place
between 1975 and 2075: see *Ecologist* 1972: 60.
[46] Schumacher 1973: 289.

system of experts to operate them. *Small is Beautiful* was, above all, a critique of Western economics and of large-scale thinking. It warned against a continuation of existing practices but its solutions contrasted sharply with the mainstream approach of *Limits*. What is more, in a quiet tone Schumacher warned of the many fallacies of forecasting, model-building, and comprehensive planning, explicitly using *Limits* as an example.[47] But above all, Schumacher argued against the discourse of growth and called for technological development to go back to 'the real needs of man': 'That also means to the actual size of man. Man is small, and, therefore, small is beautiful. To go for giantism is to go for self-destruction. And what is the cost of a reorientation? We might remind ourselves that to calculate the cost of survival is perverse.'[48]

The above account of *Limits*, *Blueprint*, and *Small is Beautiful* shows the great ambiguity within the discourse of survival that was typical of environmental politics in the early 1970s. Paradoxically, both technocratic experts and radical social movements framed the environmental problematique as a matter of survival. Both radicals and experts sought to trade in reductionist science and allocate a central role to the new systemic scientific paradigms of cybernetics and ecology. Like the environmentalists, technocrats were quick to point out that the routinized state response would fall short of delivering the structural solutions that were needed. Yet whereas *Limits* formed the basis for a coalition of forces that saw a further integration and co-ordination among the dominant social powers as the logical solution, *Blueprint* and *Small is Beautiful* became the catalyst of a coalition that sought to link the ecological crisis to a much broader social critique. Together these approaches constantly challenged the legitimacy of the rather traditional concepts and practices (often simply borrowed from

[47] Schumacher 1973. This is not to say that Schumacher provided the entirely consistent alternative to *Limits*. Schumacher failed to address the question of how his world of intermediate technology could be brought about. Here a massive big world operation seems to have been in the making. It is also interesting to note that Schumacher, himself an Economic Advisor to the National Coal Board from 1950 to 1970, emphasized the importance of coal as a future energy source (Schumacher 1973: 115 ff). Clearly, *Small is Beautiful* was not aware of the environmental consequences of coal and should thus not be regarded as the complete ecological blueprint for survival. For an account on the role of Schumacher and *Small is Beautiful* see also Bramwell 1989.

[48] Schumacher 1973: 155.

the adjacent domain of public health) with which governments
sought to regulate the emerging environmental conflicts.

Environmentalism and the Critique of Modernization

With hindsight it is possible to see that this state effort of the early
1970s to treat environmental problems like any other problem
was bound to fail. On the one hand there was the critique that
argued that the environmental problematique required a much
bigger input of science and technology. On the other hand there
was the fact that a large section of the environmentalists embedded their concern in a much broader discontent with developments in capitalist consumer society.[49] The radical environmentalism
of the social movements rested on a mixture of feelings. First of
all there was a growing fear of industrial catastrophes. The increase in the scale and complexity of industrial production, in
which professional experts had to control increasingly complex
and potentially dangerous processes, was met by a growing suspicion of the expert. Professionalism, originally meant as 'one of the
major instruments for perfectibility'[50] came to be distrusted. Secondly, like the conservation movement in the USA around the turn
of the century, the environmental movement of the 1970s expressed a growing unease about the loss of moral community.
This came to the fore through its stringent anti-consumerism, its
anti-alienation attitude, its commitment to democratization, and
its protest against what Galtung called the 'colonization of the
future'. It indicated that the radical environmentalism of the 1970s
should be seen as part of a larger critique of the process of modernization and rationalization. It was, as Suzanne Berger so nicely
put it, directed 'not against the failure of the state and society to
provide for economic growth and material prosperity, but against
their all-too-considerable success in having done so, and against
the price of this success.'[51] This general mistrust was reinforced
by the fact that the more educated public now also had easier

---

[49] Although a critique of precisely the capitalist element was conspicuously absent in the studies and was generally hidden in rather vague humanist terms. A
prominent exception here was Enzensberger's influential essay *Zur Kritik der
politischen Oekologie* (Enzensberger 1973).
[50] See Rittel and Webber 1973: 158.
[51] S. Berger, quoted in Offe 1985: 847.

cognitive access to the possibility of systemic failure and could more easily understand the particular nature of systemic irrationalities.[52] Education had created the critical attitude needed to address these systemic irrationalities.

On top of all that there was a growing frustration with the perceived insensitivity of the 'new class' of experts and technocrats to the spheres of the human life-world.[53] The darker sides of the general process of professionalization that Habermas referred to as the 'colonisation of the life-world' matched the experiences at that time.[54] People felt they had become entangled in all kinds of organizational systems that were entirely beyond their control. One of the buzzwords of the new social movements was 'alienation' and environmentalists had an easy case in showing that this also applied to the ecological crisis; they spoke of the human alienation of nature and criticized the 'anthropocentric' bias in environmental discourse.[55]

It is thus easy to see that in the discourse of the grassroots movement *Limits to Growth* appeared to be a product of the system that they thought was the very core of the problem. *Limits to Growth* followed the prevailing strategy according to which an attempt is made to find the appropriate technological answers, to put more emphasis on hierarchical management, and meanwhile to continue the routinized (consumer) practices. The radical environmentalists, on the other hand, challenged, in the words of O'Riordan, 'certain features of almost every aspect of the so-called western democratic (capitalist) culture—its motives, its aspirations, its institutions, its performance, and some of its achievements'.[56] Indeed, the radical environmental movement asked for a clarification of purposes, a re-ordering of priorities, and a redefinition of key problems. *Blueprint* and *Small is Beautiful* became constitutive elements of a discourse that rejected the technocratic practices that were inherent in the dominant idea of a survival politics as suggested in *Limits to Growth*. A technocratic response applied to an environmental problem defined in naturalistic terms profoundly missed the point. The environmental problem was primarily social

---

[52] Offe 1985: 850.	[53] See Gouldner 1979.	[54] Habermas 1981.
[55] This discursive play on the established concepts of radical political discourse was typical of the radical environmental movement at the time: see Enzensberger 1973 and also Galtung's play on Habermas's colonization of the life world, above.
[56] O'Riordan 1983: 300; see also Cotgrove and Duff 1980.

and political in character. Moreover, technocracy led to the exact opposite of what it purported to do: instead of being efficient and rational, it was short-sighted and careless. Underlying its definition of efficiency was a betrayal of the future for the sake of the present, whereas its rationality was anthropocentric and not rooted in a full appreciation of the beauty of nature and organic principles that could be derived from natural processes. In this sense 'Flower-Power' was more than just a random catch phrase.

Again, the two perspectives sketched above are not meant to represent environmental discourse of the early 1970s in full. On the contrary, there was a wide spectrum of views in which the two approaches marked two important points of focus.[57] Yet for the history of the policy discourse of ecological modernization it is significant to note how these two key statements fed two different sets of actors. These groups had quite distinct social agendas, yet at the same time they had considerable overlaps in terms of their definition of the problem. Likewise we can see how they cross-borrowed concepts, phrases, and ideas. The above shows that the alternative proposals of *Blueprint* or *Small is Beautiful* had a considerable common ground with the technocratic approach of *Limits to Growth*. Obvious examples here are the widespread application of technological fixes or the usage of comprehensive planning techniques in *Blueprint*, and Schumacher's prolonged reference to the Coal Board in Britain as an example of what it meant to be well organized. Those who were seen as the great ideologues and leaders of the environmental movement were not always as anti-technocratic as some of their followers would have wanted to believe.

3.4. CHANGING STATE–SOCIETY RELATIONS: RADICALIZATION AND CONVERGENCE

Important changes occurred in the discursive formations of environmental politics during the 1970s. The prophecy of global doom receded into the background and new issues emerged that illuminated different aspects of reality and were discussed in different

[57] Hardin and Ehrlich at one end, and Bahro or Bookchin at the other, were in many respects much more radical: see Paehlke's essay on environmentalism and the ideological spectrum in Paehlke 1989.

terms. Nuclear energy was the emblematic issue that started to dominate environmental discourse from the mid-1970s onwards. There can be no doubt that this issue changed the dramaturgy of environmental politics.[58] Yet the charisma of mass demonstrations against nuclear power makes it easy to forget that the 1970s were not only the age of mass demonstrations. Whereas the radical environmental movements captured the headlines by taking the nuclear issue to the streets during the late 1970s, rather less widely reported developments took place inside the newly institutional- ized environmental expert organizations. In the relatively concealed sphere of secondary policy-making institutes—such as the OECD, the IUCN, or the UNEP—new vocabularies for environmental policy-making were being devised. Initially these two spheres were worlds apart, but from 1979 onwards we can see a convergence of these two worlds around the policy discourse of ecological modernization. How could this unlikely process of radicalization and convergence have occurred?

During the 1970s radical factions of the environmental move- ment increasingly broke with the discursive order of lobbying and interest-group politics. Initially, the environmental movement had mimicked the institutional structures of established interest-groups that had always oriented themselves towards the state. The radical factions, however, neither wanted nor were able to negotiate their demands with the state. They related to other political actors and opponents no longer in terms of negotiations, compromise, re- form, or improvement. Politics was not a matter of gradual progress to be brought about by organized pressure-group acitivity and a strategy of inclusion. The new social movements thought in terms of sharp antinomies such as yes–no, them–us, the desirable and the intolerable, etc.[59] Especially in continental Europe, the envir- onmental movement in fact constituted an independent discourse- coalition complete with alternative life-styles and new structures of organization embracing alternative communicative practices such as mass demonstrations, and separate newspapers and radio stations.

Its actions were typically supported by slogans starting with

---

[58] The anti-nuclear movement of the 1970s and 1980s received a very elaborate treatment in the academic literature. For an overview see Rüdig 1990; Dunlap *et al.* 1993.

[59] Offe 1985: 830.

'stop', 'ban', or 'freeze' and its concern was formulated in exclusionary terms: 'no part of it [could] be meaningfully sacrificed ... without negating the concern itself.'[60] After all, what was at stake was not so much the specific way in which energy was generated or bottles were handled after their economic lives, what was at stake was their very identity or even survival.[61] This did not allow for any kind of exchange or trade-off. Given this perception of the environmental problem it is hardly surprising that nuclear-power plants became the focus of the environmental protest in the course of the 1970s. Cotgrove and Duff have pointed out the deep symbolic significance of nuclear power-stations in this respect: 'centralized, technologically complex and hazardous, and reinforcing all those trends in society which environmentalists most fear and dislike—the increasing domination of experts, threatening the freedom of the individual, and reinforcing totalitarian tendencies.'[62] The symbolic potential of nuclear power precisely played into the hands of the radical movement. The nuclear issue was pushed to the forefront of attention and became the metaphor for all that was wrong with society. In many respects the characteristics of the nuclear issue made it the logical topic for a radicalization of the survival discourse of the early 1970s. The survival discourse of *Limits* had focused on the future impacts of resource scarcity for humankind. It made environmental decline into a problem that concerned all and proposed solutions that should unite us all. On the issue of nuclear power those who held a technocratic view stood in a symmetrical antagonistic relationship to the radical environmentalists.

The respective subject positionings only reinforced this antagonistic format. In hard-core technocratic circles the anxiety over the depletion of fossil fuels, reinforced by the energy crisis of 1973/4, had always been regarded with some compassion—since nuclear power would shortly become the main source of electricity anyway (a source that would be 'too cheap to meter', as they had argued in the 1950s). Hence for the technocratic expert nuclear power was not seen as a problem but was perceived precisely as the solution to the scarcity scare. The environmental movement,

---

[60] Ibid. 831.   [61] See Cohen 1985; Melucci 1985.
[62] Cotgrove and Duff 1980: 338. Hence it would be quite wrong to suggest that environmentalism was largely a politics of 'single issue negativism' as Paehlke (1989) suggests.

however, held the opposite view. Nuclear power had increasingly become the symbol of everything that was wrong with society. Nuclear power was internally related to nuclear warfare and an essential part of the 'military–industrial complex'.[63] More subtle and often disregarded was the fact that the anti-nuclear protest also was a scientific controversy about the alleged environmental and health effects of nuclear power, to which an increasing number of scientists had become sympathetic. On top of this came the outcry over the culture of secrecy and expert domination which was critically coined 'techno-fascism'. This latter element was best expressed in Robert Jungk's *Der Atomstaat* (1977) which had the telling sub-title 'From Progress into Inhumanity'.[64]

The *Total Kritik* of the anti-nuclear movement, as it became known in Germany, and the metaphorical meaning of nuclear power therein, meant that pragmatic alternatives were not sought. That would imply a denial of the fact that nuclear power stood for a much bigger problem. The practice of mass demonstrations fitted the problem perception of the movement. Mass demonstrations near existing power plants but especially at building sites of second generation nuclear installations, such as fast-breeder reactors and reprocessing plants, became the practice through which anxiety was expressed in the USA as well as in many Western European countries, most notably (West) Germany, France, Sweden, the Netherlands, and Belgium.[65] The significance of these gatherings went completely beyond the protest against nuclear power and were, in fact, themselves quasi-religious exercises in an alternative life-style. Initiatives like the *Bürgerinitiativen* in Germany and the Dutch *basisgroepen* were attempts to decolonize the life-world and as such expressed the attachment to values such as autonomy, grassroots initiative, direct democracy, and identity

[63] This discursive linkage helps to explain the fact that nuclear power generated less open controversy in Britain than in continental Europe. After all, in Britain (a nuclear military power) the Campaign for Nuclear Disarmament (CND) had taken up a similar symbolic role. Also see Rüdig (1994), who explains the lack of attention given to nuclear power in Britain, referring to the fact that Britain did not have a planned expansion of its nuclear programme during the radical 1970s, and also quite rightly mentions the role of the inquiry practice that kept the conflict off the streets.

[64] Jungk 1977. One has to be quite cautious with international extrapolations here, since the social conflict over nuclear power was in fact quite different in the various Western countries: see Rüdig 1990.

[65] For a detailed comparative study, see Flam 1994.

formation. Hence what in the discourse of the establishment appeared as a radical action against one of its policy commitments, was for the movement an exercise of the alternative order.[66] The frequent almost military operations with which national governments responded to the demonstrations only served to add further proof to the claim that nuclear power was a threat to democracy and logically resulted in the emergence of a policy-state. No wonder that the two forces constantly talked at cross purposes. Clearly, from the environmentalist's point of view at that time, there was no room for the discussion of eco-modernist notions as described in Chapter 1.

However, the nuclear issue was an emblematic issue that had its own limited life-span. Radical protest against nuclear power slowly died out between 1978 and the early 1980s and with it went the radical environmentalist critique.[67] The social movement itself, however, did not evaporate. Research indicates that structural support for the new agenda items of environmental politics remained stable or even increased,[68] and that ecology movements continued to effect public policy in Western Europe.[69] Yet as society changed so the environmental movement changed. In the early 1980s the environmental movement found new incentives, and changed its political strategies and organizational structure. The environmentalists of the 1980s were less radical, more practical, and were much more policy-oriented. The movement's emphases were no longer on alternatives for society, it started to focus on presenting practical alternatives within society instead. Technology was no longer the focus of critique but increasingly came to be seen as the discourse of solutions. A new type of knowledge became relevant. Activists should no longer be qualified to discuss the crisis-ridden nature of the capitalist system on a reasonably well-informed level, they were now valued for various sorts of expertise (such as scientific or engineering know-how, or media, management, or marketing skills).[70] Nor did the environmental movement any longer rely solely on mass meetings for

---

[66] See esp. Cohen 1985.

[67] For the USA the key event was the accident at the nuclear plant at Three Mile Island. This lead to a *de facto* moratorium on further construction of nuclear works and thus took the wind out of the sails of the radical movement: see Rüdig 1990: 3.

[68] Dalton 1988.   [69] Rüdig 1988.

[70] See e.g. Cherfas 1990; Johnston *et al.* 1988.

its political support. Instead it aimed to maximize its campaign funds by extending both its formal membership ('giro-activism') and its wider social support—something that was completely disregarded in the 1970s. To the extent that it continued to stage mass demonstrations the meaning of these gatherings changed radically. From exercises in grassroots democracy and alternative life-styles, mass mobilization became an instrumental means to back up the activities by pressure-groups and lobbyists. The environmental movement now also emphasized the importance of media presence and behind-the-scenes political lobbying, again something that was inconceivable under the old identity-oriented movement. In short, in the 1980s the environmental movement took up the role of counter-expert, illuminating alternative solutions to what were increasingly seen again as environmental problems (thus bracketing deeper institutional causes). The problem-makers of the 1970s had become the problem-solvers of the 1980s.

The Emergence of Ecological Modernization Explained

How should this change within the movement be explained? At least four reasons can be given. First of all, radical environmentalism was caught up by the economic recession of the late 1970s. In the face of the economic slowdown environmental issues suddenly lost out against the concern over inflation and mass unemployment. The basic insecurity over the economic future of national societies frustrated the validity of the discourse of selective growth. In order to maintain its social credibility environmental discourse had to find ways to reconcile economic restructuring with environmental care, or so it seemed. Of course, within the radical political discourse the argument could have been to blame the same practices for both the environmental and economic problems. This in fact obviously happened, but somehow that claim became marginalized in environmental discourse (see below). Secondly, important changes occurred within the environmental movement itself. Even with the rather reluctant professionalization of the environmental movement at that time, the activity of professionals eroded identity orientation and favoured a strategic model. Inside the NGO-élites there was a growing awareness that the radical confrontational style unnecessarily constrained the advancement of the environmental movement as a social power. 'Soft-technologists' like Amory Lovins surfaced, arguing that the resolution of many environmental problems was well within the reach of a reformist environmentalist strategy. The practice of mass demonstrations was recognized as a dead-end and NGOs started to think about alternatives.

The third factor was the emergence of other issues such as acid rain or the diminishing ozone layer, that were not necessarily as politically illuminating as nuclear power, but that none the less seemed to be a promising basis for a further extension of the social influence of the environmental movement. The practice of strategic campaigning that had been pioneered by NGOs like Friends of the Earth and the Sierra Club, suggested the tactic of exploiting the importance of emblematic issues for the general public understanding of environmental problems. Again there was a play on the symbolic, metaphorical meaning of key issues, yet this time they had to qualify on different criteria. Rather than illustrating the perverted nature of the system at large to the radical core of the counter-culture, they now had to illustrate the vast threats that various industrial practices formed to society as a whole.

The fourth factor consisted in the political fact that an alternative discourse was available. Ideas of ecological modernization had, by then, already overcome their growing pains. Work in academic circles and expert organizations now provided an alternative conceptual language and delivered concrete solutions that suggested pragmatic ways of overcoming environmental problems. It strongly suggested that for many issues solutions could indeed be found, e.g. experiments had been conducted that hinted that energy savings of up to 40 per cent were feasible at reasonable costs. Hence what came out in the 1980s was that during the 1970s environmental politics had not only been made on the streets: a far less visible, but undoubtedly essential development of environmental discourse had taken place in the domain of policy-making institutions and think-tanks. This activity originated in the first phase of political upheaval over *Limits to Growth*. Since then policy-makers and environmental experts had staged a host of conferences and produced a stream of reports in their effort to find effective ways of regulating environmental problems. Here they could take up much work that had been done before. An important example is the Pigovian analysis in economics, which drew attention to the social costs of private enterprise.[71] The original idea stemmed from the 1920s; it had been taken up again in the 1950s and was now channelled into the policy-making arena thanks to the activities of easily identifiable mediating institutions. A key role was played by secondary policy institutes such as the OECD, the UNEP, and the UN-ECE, which had started their own environmental directorates or committees in the early 1970s. They became actively involved in the study of environmental problems and design of policy instruments. From the late 1970s onwards these institutes started to produce evidence that strongly suggested that the initial legalistic governmental response to the environmental challenge produced highly unsatisfactory results.[72] Their work thus effectively undercut the legitimacy of present (national) governmental policies.[73] What is more, their activities resulted in the formulation of a coherent alternative environmental policy that promised to be both effective and efficient. They began to call for the introduction of new policy-making strategies that focused on precaution and the internalization of environmental care in economic considerations. Hence these technocratic institutes did not operate according to the traditional image of a Leviathan. Their power was much more based on their success in providing generative story-lines that appealed to many actors in the environmental domain.

With hindsight we can identify at least three different tracks along which the repositioning in environmental discourse in the late 1970s and early 1980s took place. Together they facilitated the formation of national discourse-coalitions around the ideas of ecological modernization in the decade to come.[74]

The first track is related to the publication and subsequent take up of the *World Conservation Strategy* (WCS) in 1980.[75] This report marked the emergence of a coalition in environmental politics

[71] See Weale 1992: 158 ff; Kapp 1950.

[72] Especially significant in this respect were the *State of the Environment* reports that the OECD started to issue from 1979 onwards.

[73] Reports by the OECD Environmental Directorate, the UNEP, and similar organizations put a heavy emphasis on the serious health effects of a continuation of basic features of the industrial society such as high sulphur emissions.

[74] Here we discuss the political forces that mediated concepts such as sustainability and precaution, efficiency and self-regulation. Obviously, these concepts all have much deeper intellectual roots that go back at least as far as John Stuart Mill and George Perkins Marsh.

[75] Anon. 1980.

oriented towards policy-making and explicitly arguing for a strategy of sustainable development. The report was the joint product of the moderate NGOs the International Union for the Conservation of Nature and the World Wildlife Fund together with the UNEP, and was written in collaboration with the FAO and the UNESCO. These organizations had been working hard on a worldwide survey of endangered habitats and eco-systems. They argued strongly for a strategy of sustainable development based on efficient resource utilization and considerate environmental planning. Their agenda was to promote the conservation of nature and to this purpose the WCS came up with many suggestions for environmental policy-making.

The fact that this conceptualization of new orientations in environmental politics was generally received favourably signified a shift in the environmental domain. The policy orientation of the WCS differed markedly from the confrontational environmental action typical of the campaigns against nuclear power at that time. So while the antagonistic parties initially continued fighting their battle on nuclear power, a new discourse-coalition emerged in the environmental domain, made up of moderate forces, bridging the gap between supra-national organizations, quangos, and non-governmental organizations. The WCS was a focal point for a discourse-coalition that effectively started to exploit the middle ground that had not seemed to exist during the heyday of the nuclear controversy.

The success of the WCS indicated the political potential of an environmentally sound argument argued in a reasonable manner. The favourable reception of this report (which perhaps cannot be explained without the existence of a radical movement) most certainly contributed to the frustration with the lack of results within the radical movement. At the same time, however, it offered alternative subject-positionings for environmentalists and governments alike.

A second track along which this repositioning in environmental discourse took place was linked to the activities of the OECD. The OECD Environment Committee, launched in 1970, functioned as a think-tank and mediator for ideas that sprang up in academia. The main thrust of the OECD story-line was that pollution problems mostly indicated a gross inefficiency and that the costs of pollution should be borne by the polluters (the 'polluter pays principle').[76] As a primarily economic organization the OECD put special emphasis on the relationship of economy and environment. A key moment here was the meeting of Environment Ministers of the OECD member states in May 1979. The official Declaration on Anticipatory Policies did not go much further than the suggestion of incorporating the environmental considerations at an early stage of decision-making and a plea for the use of economic and fiscal instruments instead of concentrating on legal regulatory instruments. Yet the papers give a better idea of the underlying arguments that were important in the OECD attempt to push ecological modernization to the fore. There it was typically argued that 'In the changed economic conditions greater emphasis will have to be placed on the complementarity and compatibility of environmental and economic policies.'[77] The positive-sum game format that became characteristic of ecological modernization was simply born out of necessity: for environmental policies to survive they needed to work with the grain of the time. What is more, environmental politics was not only positioned as non-contradictory to economic policy, it was also suggested as a potential instrument for economic recovery:

The impact of environmental policies on productivity and subsequently profitability will be important in a period when investments and productivity are crucial to an acceptable rate of economic growth. The possible impacts of environmental policies on inflation, employment and balance of payments will have to be taken into account in devising policies.[78]

Whatever their intentions may have been, the OECD clearly brought eco-modernist notions to the attention of governments, emphasizing their potential in terms of economic growth while underlining the need to come to an integrated approach embracing policies towards land use, energy, transport, and tourism without which anticipatory policies were likely to be a failure. The need for a change in policy discourse was further reinforced by the study *Facing the Future* published by the Interfutures group of the OECD.[79] It showed how many fuels and materials would run out if no structural changes were made within the regime of production.

The OECD's contribution to the promotion of ecological

---

[76] Of special relevance in this context is *The Costs and Benefits of Sulphur Oxide Control* (OECD 1981*b*).
[77] OECD 1980: 60–1.   [78] OECD 1980: 61.   [79] Interfutures 1979.

modernization culminated in the International Conference on Environment and Economics of June 1984.[80] At this conference, environmental experts from various backgrounds met with governments to discuss the challenge of ecological modernization: the benefits of environmental policies, environment and industrial innovation, effectiveness and efficiency in environmental care, and the role of economic instruments. The major conclusion of the conference was that 'the environment and the economy, if properly managed, are mutually reinforcing; and are supportive of and supported by technological innovation.'[81]

The third track that helped to bring about this redefinition of the environmental conflict were the debates within the United Nations commissions on issues of development, safety, and environment. It is undisputed that the Brundtland Report, *Our Common Future* (1987), is to be seen as a sequel to the Brandt Report *North–South: A Programme for Survival* (1980) and *Common Crisis* (1983) and the report *Common Security* (1982) of the Palme Commission.[82] This sequence of UN reports signifies a continued concern with the need for increased multilateral co-operation which was strongly inspired by Western European social-democratic ideas.[83] So as in the OECD story-line we see an emphasis on the need to look at economic and environmental issues as essentially intertwined. Yet this time the background of the connection between environmental issues and economic issues of development was not rooted in efficiency concerns but stemmed from the social-democratic background of this particular brand of UN work. The UN commissions problematized the dark side of the post-war boom in production and Western welfare and in fact tried to elaborate (or keep alive) a discourse of shared global problems stemming from a perspective of solidarity. This elaborated on the state-oriented social-democratic tradition that had been outlined by the social-democratic Keynesian economists such as Galbraith and

---

[80] Interestingly, the conference was chaired by the Dutch Minister for the Environment, Dr Pieter Winsemius; see also Ch. 5.

[81] OECD 1985: 10.

[82] Independent Commission on International Development Issues (Brandt Commission) 1980, 1983; Independent Commission on Disarmament and Security Issues (Palme Commission) 1982. See also Redclift 1987. For a discussion of these reports see Ekins 1992, Ch. 2.

[83] Note that the chairpersons of the subsequent commission were all leaders of Western European social-democratic parties.

Tinbergen.[84] As such the WCED most definitely had a critical edge. It explicitly aimed to get environmental issues out of the periphery of politics (as conservation issues) and sought to link them to core—i.e. economic—concerns. Likewise it aimed to resist the dichotomy between environment and development which had been the obstacle that split North and South at the 1972 Stockholm Conference. Furthermore, Redclift notes that originally the WCED had no great belief in free-market solutions,[85] and saw a global planning approach as something that would contain the workings of the free market.

3.5. CONCLUSION: THE DISCURSIVE PARADOX

This chapter does not claim to be a history of environmental discourse but is an inquiry into the historical roots of ecological modernization. It shows the three tracks that initially propelled ecological modernization and illuminates how ecological modernization is the product of a coalition of forces in which the individual actors all had their specific interests and orientations yet found one another via a common vocabulary, and a common way of framing the environmental problematique. It indicates that the emergence of the policy discourse of ecological modernization is not to be attributed to the success or power of one particular group. Ecological modernization is the unlikely product of an argumentative interplay between several social forces that in the mid-1970s still showed radically converging ideas about the nature of the environmental problem. To be sure, ecological modernization is not all there is in environmental discourse. Yet ecological modernization is, according to the central thesis of this book, the dominant way of conceptualizing environmental matters in terms of policy-making.

This chapter has also shown that the idea of making sense of the ecological crisis in terms of what I call ecological modernization can already be found in an embryonic form in the early documents such as *Limits to Growth* or *Blueprint for Survival*.

---

[84] See Galbraith 1967; Tinbergen 1977. Interestingly, the latter author's work on the New International Order had generated interest from the Club of Rome which even led to the commissioning of the 1977 report.
[85] See Redclift 1987: 13.

Indeed, many of the ideas that were constitutive of ecological modernization have been under discussion in different places since the mid-1960s. The formation of a coherent policy discourse, however, the bringing together of these elements, was rooted in the activities of secondary policy-making organizations like the OECD, the UNEP, in the UN Commissions on environment, safety, and development, and in the intellectual work of moderate environmental NGOs such as the IUCN and the WWF.

To be sure, it was not until the mid-1980s that eco-modernist discourse really came to be a force that could effectively challenge the prevailing conceptualization of environmental problems as incidental problems that required *ad hoc* solutions. Yet from the mid-1980s onwards the regulatory failure has increasingly been recognized and accepted by actors working on environmental policy (even if only in terms of the missed chances of a modernization approach for economic development). Of course, some countries (and some departments) are markedly less inclined to suggest that the environmental problematique calls for a fundamental shift in the style and axioms of policy-making. Yet, even if the various ministries do not immediately change their procedures, we can see how from the mid-1980s onwards ecological modernization rapidly conquered the discursive space in the environmental domain and came to be seen as the most legitimate way of conceptualizing and discussing the environment as a policy-making problem. From then on, the wheels of public concern, governmental politics, and policy-making geared into one another to produce a new consensus on how to conceptualize the environmental problem, its roots, and its solutions, which was first symbolized by the global endorsement of the report *Our Common Future* in the late-1980s, and then by the Earth Summit in Rio in 1992.

The fact that ecological modernization should be seen as the product of some very distinct social forces also has repercussions for its potential to redirect the process of societal modernization. As we have seen in Chapter 1, ecological modernization is essentially an efficiency-oriented approach to the environment. This is what made it possible for ecological modernization to become the dominant discourse within the environmental domain. The positive-sum game format took away many of the objections governments might have had to a new approach of environmental regulation. What the account above shows, however, is that the original concerns of the environmental movement went precisely against the grain of that efficiency approach. Modern environmentalism emerged as an element of the counter-culture of the 1960s and this counter-culture was above all a critique of many of the technocratic institutional arrangements that are now associated with ecological modernization. Similarly, the radical anti-nuclear protests of the 1970s not only signified a politicization of the environmental debate, they also questioned the rationality of the idea of social progress. The environmental movement, in other words, has a history of reflexivity, of calling for a debate on the direction of the modernization of society. It not only argued for a resolution of the problem of physical degradation, but has always combined this demand with a critique of the institutional arrangements that were implicated in bringing these effects about and a warning of the perverse effect of a swift technocratic response. That is to say, in its best and most interesting moments the environmental movement has been concerned about the conditions under which the requirements of environmental regulation (discursive closure, social accommodation, and problem closure) were to be achieved.

The question that looms large is the extent to which the environmental movement has been able to perpetuate this reflexivity once it decided to argue its case in the appealing terms of ecological modernization. As Davies and Harré have argued, 'the taking up of one discursive practice or another . . . shapes the knowing or telling we can do.'[86] This chapter shows that the choice of ecological modernization was partly made for strategic purposes. In the early 1980s, after the disillusion of the antagonistic debate on nuclear power, the environmental movement started to argue its case in terms set by the government. Its aim was to be seen as the right kind of people, as realistic, responsible, and professional, avoiding being positioned as romanticist dreamers. While some may argue that this basically signified the environmental movement's coming of age, it is also obvious that the new discursive order imposed new limits on what could be said meaningfully. This is the discursive paradox of the new environmentalism.[87] As everyone agreed on the need to come to terms with the ecological crisis it became increasingly hard for radical groups to control the

---

[86] Davies and Harré 1990: 59.   [87] See Hajer 1990.

definition of the issues and to argue their case for a reflexive type of ecological modernization. It is obvious that environmentalists were confronted with the disciplinary mechanisms of the discourse of ecological modernization. The relationship with governments, as Paehlke and Torgerson argue, 'works smoothly only if those seeking favours are uniformly professional and responsible—if they speak the proper language of precision and instrumentality while standing ready to make the trade-offs necessary for compromise solutions.'[88]

In the 1970s the new social movements dared to bring in moral considerations and hold these against any economic calculation of cost or scientific calculation of risk, but in the 1980s this could discredit their own subject-positioning as expert. By the mere choice of vocabulary the social movements effectively restricted their own possibilities of arguing their moral case. Previously they had peeked through the aura of expertise and had escaped the modernist mythology of progress in which high-tech developments like nuclear power were embedded. With the emergence of eco-modernist discourse-coalitions this game of mutual positioning started anew.

[88] Paehlke and Torgerson 1990: 290.

4

Accumulating Knowledge, Accumulating Pollution? Ecological Modernization in the United Kingdom

4.1. INTRODUCTION

Nature has become the favourite looking-glass of modern society. We look in the mirror of nature and reflect on the merits of society: how well are we doing? The conclusions we draw differ markedly between periods and among nations. Yet it is not simply the processes of environmental change that cause these differences. What we see also reflects of the preoccupations of different periods and of different places. That is what environmental discourse reveals and here the discourses on mundane matters like pollution and waste are among the most telling. Pollution and waste are not glamorous topics but have a great social significance. Pollution is the negative that defines the positive: Douglas defined pollution as 'disorder' that legitimizes the existence of certain institutional arrangements that seek to free societies from further 'contamination', and raises questions about the legitimacy of societal institutions if they fail to perform well and produce welfare and well-being rather than pollution. It is therefore often illuminating to look at what phenomena are defined as pollution and especially how the definition of what is seen as pollution differs over time.

In this chapter we focus on the historical career of acid rain, one of the emblematic issues of the 1980s in the United Kingdom. Drawing on this case we will analyse the development of environmental discourse in Britain between 1972 and 1990 to see how eco-modernist thought affected policy-making discourse and practices.

The pollution of air and water traditionally received relatively little political attention in the United Kingdom. Pollution was a largely administrative affair while countryside conservation and the protection of wildlife could rejoice in a much greater political attention. Air pollution never was a prominent political issue until after the notorious London fog of December 1952.[1] The subsequent Clean Air Acts of 1956 are often seen as the beginning of serious attention for the air pollution issue which continues to the present. An alternative would hold precisely the opposite. It would claim that contemporary air pollution problems are of a fundamentally different quality and that the 1956 Clean Air Acts concluded the traditional era. To be sure, it would accept the argument that the 1956 Clean Air Acts led to an institutionalization of important new practices such as regular measurement of emissions and a tighter control of pollution at source. Yet it would disagree that the existing institutional arrangements can be seen as a model for regulation. On the contrary, it would see these arrangements precisely as part of the problem. This interpretive difference constitutes a central cleavage in recent British air pollution discourse. British society can be seen looking in the mirror of nature but, surprisingly, there are two widely differing ideas about what we see.

One of the striking omissions in the academic literature on British environmental politics is that this interpretive difference is rarely discussed. The literature on the acid-rain controversy that dominated the better part of the 1980s and fulfilled a key role in the debate on ecological modernization, starts from strongly realist assumptions. The question raised is why action on acid rain was slow in coming. Yet the fact that acid rain is itself a rather arbitrary social construct of the much broader problem of air pollution is left unaddressed.[2] Using the tools of discourse analysis we can reconstruct two different discourse-coalitions that interpret reality according to rather distinct story-lines. The British debate over acid rain features two distinct story-lines, one traditional-pragmatist and another based on eco-modernist principles. These two story-lines mobilized two different discourse-coalitions that stretched from politics to science, from regulators to NGOs, and from journalists to academics. The reconstruction of the dynamics of these discourse-coalitions explains how ecological modernization

---

[1] Notwithstanding the fact that historical chronicles often go back as far as 1864, 1661, or even 1273 and the British Alkali Inspectorate dates back to 1864. For the best accounts of the history of air-pollution control, see MacLeod 1965; Ashby and Anderson 1981.

[2] See Boehmer-Christiansen and Skea 1991 and Park 1986.

was taken up in Great Britain and how the problem of acid rain was regulated.

4.2. THE CAREER OF ACID RAIN 1972–1989

To be able to present a useful discourse analysis of the acid rain controversy we will first have to present the familiar sort of rundown of the most important moments in the career of the construct 'acid rain'. Acid rain entered contemporary British air pollution discourse when the Swedish Government put the issue on the international agenda at the 1972 UN conference at Stockholm. The Swedes were concerned that air pollution from Europe, but especially from the UK, might be the cause of the acidification of the natural environment in Scandinavia. Their intervention came at an unfortunate moment.[3] The emblem of resource-scarcity dominated not only environmental discourse but rated among the most prominent general political issues in 1972. The energy crisis gave people an idea of what resource scarcity might mean. Precisely at a time when the concern over energy dependency was on everybody's lips, the acid rain story discredited coal, Britain's main domestic source of energy. Furthermore, the publication of the findings of the National Survey of Air Pollution in August 1972 had indicated a 30 per cent fall in SO_2 emissions and a significant improvement in sulphur dioxide concentration at ground level between 1960 and 1970.[4] It strongly reinforced the official post-1956 story-line that suggested that air pollution was an issue on its way out.[5] On top of that, the acid-rain story suggested that air pollution had detrimental effects on Swedish lakes, at a time when the National Society for Clean Air, a powerful NGO, had just started to campaign against a new kind of urban air pollution:

---

[3] Statements about the relative importance of various pollution issues are primarily based on a content analysis of *The Times, Nature,* and *New Scientist.*

[4] Warren Spring Laboratory 1972.

[5] It is questionable whether the figures should have been read like this. The Survey also revealed that the fall in SO_2 was the result of an impressive decrease of domestic output (from 0.86 m. tonnes in 1950 to 0.50 in 1970) whilst output by industry had remained about the same. The output of SO_2 by power-stations had gone up from 0.95 million tonnes to 2.65 in 1970. Over the period 1950 to 1970 we can see a dramatic increase rather than a 30% decrease. See Anon. 1972: 366.

photochemical smog. Photochemical smog was an issue that matched the routinized institutional bias of air pollution organizations towards the urban realm and public health, and quickly came to dominate air pollution debates at the expense of the acid rain story.

Hence acid rain did not exactly hit the political agenda after the Stockholm conference, let alone being recognized as a programmatic issue calling for major institutional change. Indeed, due to lack of public attention acid rain was relegated to the institutional agenda of British policy-makers and scientists that had to prepare an official response to the Swedish allegations. In all, the 1970s were a period in which research projects gradually revealed the extent of the problem while Scandinavian governments constantly pressed for swift action. In this climate of contradictory claims the UK Review Group on Acid Rain was set up in 1980 to examine whether there was acid deposition on British soil.[6]

However, while the Review Group was studying the possible effects of acid rain in Britain, the boundaries of environmental politics were being substantially redrawn. In 1979 the Conservatives came to power. The new Prime Minister, Margaret Thatcher, had an unequivocal political programme for the economic restructuring of Britain according to neo-liberal prescriptions. The central concern was to 'roll back the state' and improve the general climate for business, which certainly did not leave space for an active environmental policy. This also affected the realm of air pollution regulation. In his first general report, the new Secretary of State of the Environment, Michael Heseltine, made a firm statement about the 'steady improvements' in air pollution control. It also heralded the return of the 'tall stacks' as a success story despite the fact that an OECD study had demonstrated the relationship with acid rain:[7]

Sulphur dioxide emissions . . . fell [after 1970] by about 16 per cent and recently seem to have been roughly stable at a new low level, while

---

[6] The Review Group published its final report (containing mainly deposition levels) in December 1983: see UK Review Group on Acid Rain 1983.

[7] See OECD 1977. In fact a British government spokesman, Dr L. E. Reed, then secretary of the DoE Discussion Group on acid rain, had admitted in public in June 1976 that SO_2 emissions from Britain's high chimneys did indeed reach Scandinavia at the international conference on acid rain at Telemark (Norway): see Reed 1976.

average urban concentrations have fallen by about 50 per cent since the early 1960s: the difference between the pattern in emissions and concentrations is the result of more effective means of dispersal, e.g. higher chimneys.[8]

In actual fact the whole acid rain issue and the international dimension of air pollution were omitted from the report.

In June 1982 a group of scientists of the Institute of Terrestrial Ecology (ITE) published an article in *Nature* that showed that acidity levels in Britain were comparable to those causing fish death in Scandinavia.[9] They published their results just before the second Stockholm conference that was meant to examine progress at the tenth anniversary of the UN conference of 1972. It is not very likely that these figures published in *Nature* would, by themselves, have generated much public attention if they had not been paralleled by the sudden conversion of the German government. At Stockholm the German representatives, who had always resisted Scandinavian pressure, issued a dramatic account on the scale of *Waldsterben* in West Germany and announced an expensive crash programme to cut their SO_2 emissions. With the German conversion Britain lost an important ally which only reinforced the Scandinavian pressure on the British government.

In Britain the prominence of the lead issue masked concern over acid rain at the time.[10] It was not until the lead issue was resolved in April 1983 that campaigners, journalists, politicians, and government officials really became sensitive to the issue of acid rain.[11]

---

[8] DoE 1979: 3.   [9] Fowler *et al.* 1982.

[10] In April 1983 the Royal Commission on Environmental Pollution published its Ninth Report, *Lead in the Environment*. The government used this as a triggering device to change its policy, announcing that it would follow the recommendations of the Report.

[11] This can be illustrated by a calculation of the entries in *The Times* Index relating to lead in petrol and acid rain. Both are compound categories comprising more than one heading. Lead in petrol was an important issue in the early 1970s (with 22 entries in 1972) but subsequently lost importance (1978 even indicated no entries on lead in petrol). In 1980 it suddenly scored 27 entries and in 1982 even 51. After 1982 attention fell sharply (4 entries in 1984; 15 in 1985; 4 in 1986). Acid rain, on the other hand, scored between 1 and 5 entries annually during the 1970s but started to rise after 1981. In 1982 it scored 15 entries; in 1983 18 and in 1984 in peaked at 67 entries. After that acid rain dropped to 42 in 1985, 38 in 1986, and 21 in 1987.

In May 1983 acid rain gained a dominant position in the discursive space of environmental politics. It kept a prominent position until at least 1987. The NGO Friends of the Earth started to campaign on the issue in 1983 and journalists were drawn to the German pictures of dying trees and started to report on the concerns abroad. The Government fended off its critics maintaining that 'the politics of acid rain have run ahead of the science'.[12] Yet under influence of foreign examples concern over acid rain quickly became respectable. In February 1984 the Royal Commission on Environmental Pollution identified acid rain as one of the most important pollution issues of the present time.[13] The Royal Commission on Environmental Pollution had only been sworn in in 1971 but had established itself as an authoritative institution, thanks to a range of generally approved reports and chairmanship held by various respected scientists. Other reports on acid rain followed, yet the single most important decision was undoubtedly the decision of the House of Commons Select Committee for the Environment to hold an inquiry into the acid rain problem. The inquiry hearings were held in the period from May to July 1984 and became one of the high points in the British controversy.

The report of the all-party Select Committee was published in September 1984. It did not accept the discursive confines within which the government had sought to define the acid rain issue and identified acid rain as 'one of the major environmental hazards faced by the industrial world today'. It argued that 'enough is now known' and urged the government to join the 30 per cent Club (which would require a 30 per cent reduction in the SO_2 levels of 1980 by 1993). It also recommended that the CEGB should retrofit its power stations with expensive Flue Gas Desulphurization equipment (FGD) to achieve a 60 per cent reduction in SO_2 emissions by 1995.[14] In December 1984 the government responded with a proposal for a reduction in SO_2 emissions of 30 per cent by the end of the 1990s but it did not accept the necessity of

[12] *The Times*, 7 May 1983: 8.

[13] Royal Commission on Environmental Pollution 1984: 147.

[14] House of Commons Environment Committee, 1984 (hereafter HoC Environment Committee); the Select Committee on the European Communities of the House of Lords had also argued that two power-stations should be fitted with FGD: see their 22nd report (1984).

installing expensive FGD scrubbers. This was regarded as too costly.[15]

After 1984 the acid rain controversy became somewhat repetitive and predictable. Pressure built up but new arguments were scarce. This lasted until September 1986, when the government announced its decision to install FGD to all new coal-fired power-stations. 'New scientific evidence' had persuaded the government. The government also announced plans to retrofit three of its twelve large coal-fired power-stations with FGD at a total cost of approximately £600m. The Norwegian Ministry for the Environment was not impressed. It described the measure as a 'slap in the face' and received support from the Swedish government and Sir Hugh Rossi, the chairman of the Select Committee.[16] In May 1987 the CEGB announced it was to spend £170m. to fit twelve power-stations with new 'low NO_x burners' to reduce its NO_x emission by 30 per cent.

In the final years of the decade acid rain was a far less prominent issue. The 1986 decision had taken the sting out of the tail and NGOs had put their campaigning money and manpower elsewhere. Acid rain now had to share discursive space with new issues like the greenhouse effect and the diminishing ozone layer. In January 1988 the Environment Committee of the House of Commons held another inquiry but this time its subject was defined as Air Pollution.[17] The arbitrary confinement of their subject-matter was thus to a large extent corrected. Indeed, in the 1988 report suddenly a great number of air-pollution issues were problematized. The need to look at environmental problems in the round was further reinforced by the Brundtland Report, which was endorsed by the Government in July 1988.[18] Acid rain was one of the illustrations of the troublesome environmental condition. Yet at the same time the government was particularly keen to show that sustainable development did not call for a break with past commitments. The tone of the reply was that 'much still

[15] DoE 1984a. The discrepancy was bigger than immediately meets the eye, since at that time it was believed that this percentage would be achieved as a result of the then apparent tendency of SO_2 emissions to fall and therefore no governmental action would be needed to meet this goal.

[16] Walgate 1986: 191; *The Times*, 11 Sept. 1986: 9; *The Times*, 15 Sept. 1986: 2.

[17] HoC Environment Committee 1988.

[18] DoE 1988a.

remains to be done', refinements and improvements were to be made, and 'where appropriate the government will take further precautionary measures.'[19] The report explicitly stated that 'we are not convinced of the need for changes to the machinery of the UK government.'[20]

In June 1988 the long-lasting negotiation over the EC Large Combustion Plant Directive (that had started in 1983) finally came to a close. It resulted in an agreement which gave Britain more lenient terms to make up for the high expenditure of FGD installations. The European Council of Ministers agreed on a UK SO_2 emission target of 60 per cent of 1980 levels before 2003. The agreement reflected the old biases in British air pollution regulation: the emphasis was on finding pragmatic technological solutions combined with a belief that emissions in themselves are legitimate and a resistance to any categorical regulation.

Right at the end of the decade traditional-pragmatic institutional commitments came under fire as an unintended consequence of the government's plan to privatize the electricity industry. Quite suddenly this opened up some regulatory routines. The plans for the implementation of emission reductions had always started from the assumption that FGD was the only instrument that could be employed to meet SO_2 reduction targets. Yet the new split between regulator and regulated created new options for SO_2 reductions. The import of low sulphur (and low cost) coal and the employment of natural gas became economically attractive new options.

4.3. ACID RAIN AS TEXT

A run-down of facts and developments can only give a superficial idea of the regulation of acid rain. The British acid rain controversy involved many different actors, each with their own preoccupations and interests, and each with their own particular contributions and insights. Yet basically the argument can be represented as a competition between two distinct discourse-coalitions that were organized around two different story-lines:

[19] DoE 1988*a*: 8.   [20] DoE 1988*a*: 55.

one drawing on the categories of traditional pragmatism, and the other using the concepts of ecological modernization.[21]

The Traditional-Pragmatist Story-Line

The traditional-pragmatist story line has its origins in nineteenth-century air pollution discourse. The most eloquent and outspoken narrator of this story-line was the Central Electricity Generating Board (CEGB). In actual fact even the Government more or less echoed the detailed arguments that were presented by the CEGB, adding a more explicitly political-administrative dimension. This story-line evolved around a natural science conceptualization of the problem. Acid rain was framed as a question for research. It had to be established whether there was genuine environmental damage which could be attributed to sulphur emissions produced by CEGB power-stations. If this could be proved, and if FGD would be shown to be environmentally effective as well as the most cost-effective solution, FGD should be installed. Acid rain certainly looked like a serious pollution issue but it was not seen as an anomaly to the institutionalized way of dealing with pollution.

This commitment to science was substantiated by research by the CEGB itself. The CEGB had started its own research into acid rain in the late 1960s and despite the fact that it housed the single most important research group on the subject, in the early 1980s it still argued that the available evidence was 'anecdotal and intuitive' and that there was a need for 'proper' research.[22] It argued that there was no real scientific understanding let alone a consensus on the mechanisms involved in lake acidification. In 1984 Lord Marshall (then Sir Walter), chairman of the CEGB and member of the Royal Society, argued in his capacity as manager: 'I simply do not accept any of the scientific arguments I have yet seen.'[23] Here

[21] This section will largely be a synchronic analysis comparing two story-lines. Obviously, in actual fact they underwent substantial changes over time. This diachronic analysis will receive more attention in subsequent sections.

[22] Dr G. Howells (interview), Chester 1986; Marshall in Pearce 1986: 23. It should be emphasized that this interpretation was not uncontroversial within the scientific world: see e.g. K. R. Ashby 1982 and the evidence presented to the Environment Select Committee by the NERC (HoC Environment Committee, Minutes of Evidence (MoE) 1984).

[23] HoC Environment Committee (MoE) 1984: 20.

the CEGB management consciously presented its own science-based argument in contrast to the position of the Scandinavian and German governments. These governments had argued in the early 1980s that no further research was needed and suggested that the state of knowledge at that time justified immediate action. The CEGB thus positioned foreign governments as emotional and irresponsible and itself as rational and scientific.

The main means by which this commitment to science was substantiated was the reference to the SWAP research project. The SWAP project (Surface Waters Acidification Programme) was launched by the CEGB while public pressure was building up in September 1983. It was a £5m. research study into the acidification of fresh water funded by the CEGB and the British Coal Board. The SWAP project was to be conducted by the Royal Society jointly with the Royal Swedish Academy and the Norwegian Academy of Science and Letters.

At that time the CEGB researchers were aware that acid precipitation might be responsible for the acidification of some lakes in southern Scandinavia but they argued that more research was needed to establish whether a reduction of SO_2 emissions from Britain would have any effect on the fishery status of these lakes. Frequent reference was made to the voluntary commitment of the CEGB to the outcomes of SWAP, whatever the findings might be. It 'would get on with the job, whatever the costs'.[24] Even more prominent was the constant reference to the involvement of the Royal Society of London and the Swedish and Norwegian Royal Academies of Science. They were put forward as the 'most prestigious scientific academies in the entire world'.[25]

This commitment to science was the primary basis for the credibility claim of the traditional-pragmatist story-line. The underlying motive was that action should be ecologically effective and, as far as possible, cost-effective and that therefore an adequate understanding of what was controlling the chemical balance in nature was needed.[26] This scientific core was also legitimized in moral terms. It was argued that 'Since electricity is essential for

[24] See e.g. Marshall, reported in Walgate 1984*b*: 94.
[25] Lord Marshall, in HoC Environment Committee (MoE CEGB) 1984: 36.
[26] On cost-effectiveness see: HoC Environment Committee (MoE) 1984: 5; on the chemical balance see statement by Lord Marshall at the launch of the SWAP-project, 5 Sept. 1983, (CEGB 1983: 3).

everyone, this [higher price, MAH] effectively lowers everyone's standard of living and it would be tragic to do this without understanding exactly what we are accomplishing'.[27] Scientific understanding was thus defined as the only truly rational basis for (political) action. The story-line further implied that if SO_2 emissions could not be proved to be harmful to the environment they should continue to be emitted.[28] This reinforced the established pragmatic approach to pollution.

The politicians who drew on the traditional-pragmatist story-line used the commitment to science in the context of a strict utilitarian approach to environmental decision-making. The official line of the government was to insist that it would be willing to act if action would be 'environmentally effective and economically feasible'.[29] It argued that it

does not believe that the very substantial expenditure (running into hundreds of millions of pounds) which would be required to install flue-gas desulphurisation plant at existing power stations can be justified while scientific knowledge is developing and the environmental benefit remains uncertain.[30]

The government positioned its emphasis on the need for a better chemical understanding against the call from its critics to make 'heroic efforts'. As William Waldegrave, then Minister at the DoE, said in 1984: 'We see no point in making heroic efforts, at great cost, to control one out of many factors unless there is a reasonable expectation that such control will lead to real improvement in the environment'.[31] The British government argued that politicians rushed (EC) or were in danger of rushing (UK) to conclusions on the basis of fallacious data and arguments. The politicians who defended the government's stand legitimized their position by emphasizing their accountability to the electorate: if a lot of money would be wasted they would be punished by the electorate.[32] The

---

[27] CEGB 1983: 4.   [28] Especially evident in Chester 1989.

[29] Statement by the DoE at the Munich Conference, June 1984 (DoE 1984c).

[30] DoE 1984d: 3. Similar statements are reported by minister Giles Shaw at the 1982 Stockholm conference (see Pearce 1982: 80) and Secretary of State Patrick Jenkin in March 1984 (see DoE 1984a; Park 1986: 222).

[31] Speech to closing session of the Munich Conference on acid rain, 27 June 1984 (DoE 1984c). The contradictory statements made by Waldegrave in the context of the acid-rain controversy will be discussed below.

[32] Waldegrave in HoC Environment Committee (MoE) 1984: 326.

traditional-pragmatist story-line thus defined the meaning of responsibility and justice in relation to the national taxpayer. It even raised the question whether it was legitimate to use taxpayers' money to solve foreign problems.

An essential element in the traditional-pragmatist story-line was the attempt to enhance the legitimacy of the pragmatist position by the continuous reference to the success of its pragmatic and science-based approach in the past. The overall assessment of its own effectiveness in the past was that Britain had a 'proud record' in air pollution control.[33] The tall stacks policy had resolved the problem of urban air pollution and the fact that SO_2 emission had fallen by about 30 per cent since the early 1970s was used to add weight to this assertion.[34]

In line with the traditional-pragmatist approach, government actors admitted that SO_2 might cause some damage to the natural environment but emphasized that it was only one component in a complex situation alongside natural factors like climate and specific environmental parameters such as the condition of soil, air, and water.[35]

Like the CEGB, the Government positioned itself as the more rational actor compared to governments abroad. The target of the 30 per cent Club, for instance, was perceived to be an irrational and arbitrary basis for action which would give Britain an unfair disadvantage. Furthermore, the fact that Britain had a favourable geographical position could hardly be used against it.

To be able to uphold the traditional-pragmatist approach it was important to approach acid rain as a pollution incident like all others, albeit of a serious nature. Hence special features were played down. This happened in all three key areas of concern: damage to buildings, forest dieback, and the acidification of the Scandinavian environment. The damage to 'our historic buildings', it was argued, was not new but had been going on for generations; the concern over forest dieback was a case of 'false alarm' which probably had more to do with altitude, drought, and cold winters

[33] DoE 1979: 3; DoE 1982: iii; DoE 1984d: 3, 8, 11.
[34] 'About 30%', press notice by Patrick Jenkin, 22 Mar. 1984 (DoE 1984a); '37%' DoE at Munich Conference, June 1984 (DoE 1984c); 'nearly 40%', DoE in reply to HoC Environment Committee (DoE 1984d: 3).
[35] 'Damage to the natural environment' (Waldegrave in HoC Environment Committee (MoE) 1984: 331); 'complex situation' (DoE HoC Environment Committee (MoE) 1984: 67); 'climate, soil, air and water' (DoE 1984d: 4).

than air pollution; the damage to trees in the UK bore 'only a superficial resemblance' to that of the trees in Germany, whilst fresh-water acidification was 'not a simple story' and required more research.[36]

Various actors employed the traditional-pragmatist story-line in the acid rain controversy. Organizations such as the Confederation of British Industry and the Forestry Commission all structured their arguments around this story-line. Yet with the exception of the Forestry Commission, whose argument will be discussed in detail below, they were less eloquent and did not add fundamentally new elements to the comprehensive story-line that was produced and reproduced by the CEGB. The arguments sounded all too familiar: the CBI defended the interest of British industry to keep electricity prices down, argued that costly measures were premature and unwise, and suggested that FGD would not be a commercially viable option. It referred to scientific evidence arguing that there was no immediate need for action. The Society of Motor Manufacturers and Traders Ltd. argued that the problem was not well understood, that the contribution of motor cars was limited, and that we were on the brink of a new combustion technology anyway. The Watt Committee on Energy[37] criticized the call for action when the scientific community was not sure about the effectiveness of action. It also maintained that if acid rain caused problems, liming, certainly one of the most paradigmatic examples of a pragmatist approach, would be the appropriate solution.[38]

As the government and the CEGB announced the decision to

[36] Damage to buildings (DoE in HoC Environment Committee (MoE) 1984: 106); forest dieback (DoE in HoC Environment Committee (MoE) 1984: 17, 100; DoE 1984d: 8); fresh-water acidification (DoE 1984d: 7).

[37] The Watt Committee on Energy represented 60 British professional organizations, mainly from the field of engineering and industry. The chairman of its Working Group on Acid Rain was professor Kenneth Mellanby. This working group later received research grants from the Department of Energy (see Walgate 1984a: 535).

[38] CBI in HoC Environment Committee (MoE) 1984: 191, 200; Motor Industry in HoC Environment Committee (MoE) 1984: 140–2; Watt Committee 1984. There was an even more extreme example of the pragmatist approach on technological solutions than liming. Scientists from the Meteorological Office suggested that, since a high proportion of acid rain fell within a few rainy days, one could use meteorological forecasts to predict these days and than ask power-stations to switch to low sulphur fuels (reported by the DoE in HoC Environment Committee (MoE) 1984: 99).

install FGD in September 1986, this was seen as a confirmation of the traditional-pragmatist approach: new evidence had shown that action was needed and the research had established it would be both environmentally effective and economically feasible.

The Eco-Modernist Story-Line on Acid Rain

The traditional-pragmatist story-line on acid rain was fiercely attacked by actors from both within parliament and outside. The three main opposing actors were the Environment Select Committee of the House of Commons, the Royal Commission on Environmental Pollution, and the environmentalist NGO Friends of the Earth UK (FoE). Drawing heavily on the newly emerging discourse of ecological modernization they defined the acid rain issue in a different and wider context. Their eco-modernist story-line on acid rain inevitably started from the perception that acid rain was not just another issue: the Select Committee saw acid rain as 'one of the major environmental hazards faced by the industrial world today'; the Royal Commission argued that 'acid deposition is one of the most important pollution issues of the present time' whilst FoE contended that 'acid rain is already widespread in Britain.'[39] The issue was too big to be treated in the established incremental way. It was seen as a Kuhnian 'exemplar' of a new sort of environmental problem requiring major institutional change. Many of the other defining characteristics of ecological modernization can be traced in the arguments of the oppositional forces in the British acid rain controversy. This is evident in what was undoubtedly the most influential eco-modernist statement on acid rain: the Fourth Report of Environment Committee of the House of Commons.

The Fourth Report shows the eco-modernist story-line in its typical British form. It positioned itself against the government and played down the monopoly for scientific assessments of the situation. Knowledge about acid rain was imperfect but 'time was running out'.[40] 'Enough is now known', it was argued from 1982 onwards, to justify the spending on curative measures. It was also thought to be obvious what had to be done: the report followed the Swedish government that saw the CEGB power-stations as the

[39] HoC Environment Committee 1984: 1; Royal Commission 1984: 147; FoE in HoC Environment Committee (MoE) 1984: 39.
[40] HoC Environment Committee 1984: xi.

main culprit, and so retrofitting power-stations with Flue Gas Desulphurisation (FGD) was put forward as the 'proven techno-logy' to cure the problem. Furthermore the HoC Environment Committee attacked the responsibility claim on the part of the CEGB and the government. The HoC Environment Committee drew attention to the fact that the CEGB spent £1.5m. on the environ-mental effects of acid rain but just £200,000 on FGD-related re-search.[41] Contrary to the traditional-pragmatist story-line, the HoC Environment Committee emphasized the negative economic effects of acid rain. It used a cost–benefit analysis carried out by the consultancy Environmental Resources Ltd. to argue its case.

Instead of emphasizing the 'proud record' in national air pollu-tion abatement, the eco-modernist story-line positioned the UK in international perspective. It underscored the fact that the UK still was the largest producer of SO_2 in Western Europe.[42] In contrast to the traditional-pragmatist story-line that positioned acid rain with respect to the overall decline in SO_2 emissions, the eco-mod-ernist argument was typically illustrated with a graph that indi-cated the relative increase of SO_2 emissions due to power-stations.

Hence the emphasis on science was replaced by a responsive political approach: 'simply to plead for more research into cause and effect is but to procrastinate.'[43] This call for action was not so much a purely ethical affair but was based on the idea that issues like acid rain were substantially different from earlier pol-lution issues and hence called for a more active approach. It was not the reference to science that distinguished the eco-modernist story-line from the traditional-pragmatist story-line. The distinc-tion lay in the kind of scientific evidence that was deemed appro-priate and its connection to particular policy actions. Like the traditional-pragmatist story-line, the eco-modernist story-line was supported by scientific advice. Yet it argued that research should not be devoted to finding 'scientific proof' of the ecological phe-nomenon but to making an assessment of 'risk' involved in action and non-action. Expert bodies such as the NERC and NCC were drawn upon to legitimize the claim that the environmental risks were too high and decisions should be taken.[44]

[41] HoC Environment Committee 1984: lviii.
[42] Royal Commission 1984: 144; FoE in HoC Environment Committee (MoE) 1984: 37, 40; HoC Environment Committee 1984: xi, xiii.
[43] HoC Environment Committee 1984: xi.   [44] Ibid. xiv.

Another characteristic of the British eco-modernist acid rain story-line was the attempt to focus the attribution of blame. This was inspired by political considerations. It was argued that coal-fired power-stations were responsible for the main part of the SO_2 and NO_x emissions, and that action should therefore focus on that single source.[45] Although it also concluded that the NO_x emissions due to traffic had increased, it did not push this socially delicate aspect. And although it signalled the potential importance of ozone and the influence of the motor car, it was rather weak in its recommendation in this respect too. In other words, there was a conscious attempt to seek solutions that matched the public under-standing of the issue. This was also evident in the HoC Environment Committee support for direct symbolic action: Britain should join the 30 per cent Club immediately.

Other actors had their own little differences with the HoC Environment Committee. The Royal Commission's recommendations on acid rain, for instance, did not, in fact, really match the eco-modernist tone that prevailed in its celebrated Tenth Report (1984). Although the Royal Commission argued that it was convinced of the damage due to acid rain, and argued that FGD retrofits were the only short-term solution, it failed to make the recommendations that would match its analysis. Instead it argued that 'high priority should be given to research on acid deposition, in particular on the causes and effects, on the interaction with other pollutants, and on remedial action.' It further recommended that the CEGB should introduce on a pilot basis certain abatement options.[46] The Royal Commission also refrained from recommending that the UK join the 30 per cent Club.

The third key actor operating within the eco-modernist frame was Friends of the Earth.[47] FoE put special emphasis on the fact

[45] Ibid. lii.
[46] Royal Commission, 1984: 147. It allowed the CEGB to argue that it was making the research effort that the Royal Commission suggested (HoC Environment Committee (MoE) 1984: 6). The recommendations were also heavily criticized in a joint statement by Nordic environment ministers (Statement to the British Ambassador in Stockholmon 1 Mar. 1984). They described the recommendations of the Tenth Report as a 'serious setback'. This apparent failure to apply its own criteria will be explained in the discussion below.
[47] The fact that Friends of the Earth receives more attention here than other NGOs is simply a reflection of the co-ordinated action among the British NGOs at the time. Friends of the Earth was to lead the campaign against acid rain and apart from FoE only Greenpeace and the NSCA played an active role on this front.

that British flora and fauna had also been damaged and were threatened by further decay. FoE twice organized a survey of tree health in Britain to prove that acid rain was not just a Scandinavian problem (see below). FoE argued strongly for an electricity conservation programme, that could reduce demand by 5 per cent as an additional measure to the FGD retrofits. This would be based on the usage of more efficient appliances of various sorts. Later this sort of prevention should replace scrubbing as a policy strategy. FoE promoted this preventive strategy on the basis of the conviction that 'it is likely to be more expensive to act later rather than sooner'.[48] Like the Royal Commission, FoE referred to the UK response to the World Conservation Strategy as a useful way of looking at environmental problems (see below). FoE was strongly in favour of joining the international initiative of the 30 per cent Club. The 1986 decision was interpreted as a case of 'too little, too late'.

4.4. DISCOURSE-COALITIONS IN THE ACID-RAIN CONTROVERSY

The acid-rain controversy was far from being a debate about facts alone. It involved substantial themes like the seriousness of—or even the very existence of—an environmental crisis, the effectiveness of existing strategies of regulation, the appropriate relationship of science and policy, themes like the role of acid rain in maintaining social order, as well as questions of morality, responsibility, and justice. Yet to be able to understand the political impact of these story-lines they should be seen in the context of the argumentative game in which they acquired, sustained, and transformed their meaning. Story-lines as presented in Section 4.3 are obviously ideal-typical compositions. Only at rare moments in the controversy can we find the in-depth exchange that allowed the many facets of both perspectives to emerge. Argumentative discourse theory suggests, however, that the utterance of just an element of either of these story-lines is understood by others in the context of their implicit knowledge of a particular story-line. So

---

[48] HoC Environment Committee (MoE) 1984: 8, 42, 51.

comprehensive understandings can be reproduced in by and large symbolic exchanges of references.

We have seen how the fact that air pollution regulation could glory in historical roots in the nineteenth century was time and again used to structure the traditional-pragmatist story-line on acid rain. The moral was that our institutions did well in the past, and that they will prove their value now. This did not markedly change under the influence of the Thatcher government. The burden of proof therefore lay with the eco-modernists. They had to come up with the arguments that showed that the existing institutions would not be in the position to deal with the problem. Here the *World Conservation Strategy* (WCS), published in 1980, and especially the official British appendix to the WCS's published in 1983, proved to be important and often acknowledged sources of inspiration. We will therefore present their arguments in some detail.

The WCS was a comprehensive investigation of conservation issues, jointly published by the UNEP, the IUCN, and WWF, and indicated the growing significance of ideas and concepts of ecological modernization. To be sure, the direct effect of the original WCS on air pollution discourse was minimal: it only touched upon the acid rain issue.[49] Yet the WCS was significant for its insistence on the need to change the routinized cognitive categories within which environmental problems were conceived. Pollution could no longer be seen as legitimate, and the problem of environmental decline was too serious to be approached by a purely pragmatic response: conservation and pollution control required a coherent strategy. The WCS recommended a review of existing national legislation and recommended specific new organizational principles. This theme was picked up in the British response to the World Conservation Strategy which appeared in 1983. Although it should be emphasized that the status of *The Conservation and Development Programme for the UK* remained unclear, it reflected many of the ideal-typical features of ecological

---

[49] The WCS saw things in a broad perspective on the environmental problematique and this caused the WCS to suggest that although acid rain was a serious problem that should be resolved, it might in fact simply be the first issue to surface from the structural problem of the accumulation of gases in the atmosphere (Anon. 1980, sect. 18.8).

modernization.[50] It can be seen as the introduction of eco-modernist thought to the centre of British environmental politics.

The Conservation and Development Programme emphasized its belief in a positive-sum game solution and in the reconciliation of ecological sustainability and economic growth. In this respect it was strongly influenced by the ideas of Alvin Toffler's *Third Wave*. The report identified 'seven bridges to the future' describing seven 'sunrise' activities that could play a role in the reindustrialization of Britain in an ecologically sound way:

> Herein lies a major 'sunrise' opportunity for business: bold new approaches to conserving resources and combating pollution. The opportunities are manifold. They are already being seized by dynamic companies which have learned to turn incidental cost into profit. One multinational, which has developed a systematic programme of pollution control throughout its activities, insists that 'pollution prevention pays'.[51]

The report also introduced ecology into policy discourse although in what was a typical case of discursive contamination. Ecology featured as an appropriate scientific paradigm that was used as if it pointed 'the way to new and more rewarding lifestyles, affording greater harmony between the human species and nature, and consequently greater harmony among people.'[52]

By and large *The Conservation and Development Programme* offered a well-informed and quite critical perspective on the 'environmental challenge'. One of the key elements in its general discourse of opportunity was, for instance, the heavy emphasis on the potential of Britain's 'fifth fuel'—energy conservation, which was for instance drawn upon by FoE. It signalled the collapse of various governmental conservation schemes and the bias towards nuclear energy *vis-à-vis* renewable sources, it criticized the overall

[50] The UK response was the product of an unusual request by the Secretary of State for the Environment. He had welcomed the original WCS but instead of preparing an official reply he passed this task on to a variety of official advisory organizations and more autonomous environmental organizations who acted as sponsors and subcontracted the actual study and writing to a number of autonomous authors: see Anon. 1983a: 10.

[51] Johnson 1983: 36. Apart from the dubious credibility-reinforcing strength of the reference to 'one multinational' the report at many places refers to evidence that countries like Germany and Japan are already well under way in a process of ecological modernization.

[52] Anon. 1983a: 11, 22 ff.

'ideology of growth' that concentrated on the (profitable) stimu-lation of supply of energy instead of the reduction of demand. It did not fail to notice that the burden of this strategy would fall on those 'least able to afford it'. *The Conservation and Development Programme* was a rare example where the positive-sum game format was drawn upon to show that resolving the environmental crisis could not only help in the fight against inflation but could also help to create jobs.

The WCS did not become the dominant approach to environ-mental politics. Far from it. Yet its well-publicized ideas intro-duced new arguments into the environmental arena. The WCS further led to some vague attempts to create an eco-modernist coalition embracing actors from industry, science, and government in the context of UK-CEED. What is more, the UK reply to the WCS in the end even prompted an official governmental response in May 1986. Here the Government for the first time formally endorsed the principle of sustainable development. Yet the docu-ment disappointed those who had any dealings with the UK reply to the WCS: it hardly referred to the original document and was mainly a reiteration of the traditional British style of environmen-tal policy-making.[53] Nevertheless the WCS was an important intel-lectual source of inspiration and it caused critics of the government like Friends of the Earth to break with their strict campaigning practice to not only argue for the installation of FGD, but also to challenge the basic premises of the environmental policy of the government.

However, the argumentative perspective implies that it would be wrong to think that these story-lines were glued to specific actors. They are better seen as inherent in certain social practices in which specific social relations are defined.[54] This is a rather important claim that allows for the understanding of the initially rather puzzling fact that actors are often found expressing contra-dictory statements. This is, in fact, a recurring feature of the British acid rain controversy. Depending on the practice or argumentative

---

[53] DoE 1986.
[54] This is not a structuralist claim: actors are themselves actively taking part in reproducing and transforming practices. What is emphasized is the fact that actors enter into routinized patterns of social behaviour which include clear rules of prohibition, exclusion, and discipline. These can be contested, denied, or fought but only at a substantial intellectual and social cost.

situation in which they participate, actors can be found drawing on elements from both story-lines. An example of this relational positioning is Sir Hugh Rossi's statement in September 1986. At the high point of the controversy Rossi, one of the key actors in the proliferation of the discourse of ecological modernization in Britain, pointed out at a conference in Stockholm, that Britain 'had led the way' in curbing air pollution since the Clean Air Act and argued that in terms of NO_x Britain was 'ahead'. He thus helped to reproduce the myth of the proud record, a key element of the traditional-pragmatist story-line on acid rain. Another and, perhaps even more remarkable example is the statement by Dr Martin Holdgate, then Chief Scientist at the DoE, at the big acid rain conference at Munich in June 1984. Holdgate defended the typical pragmatist decision by Britain to postpone action, but drew on an eco-modernist line. He said: 'We seek new technologies that achieve pollution abatement as an integral component of design rather than add on costly curative devices to inherently dirty plant.'[55] This can easily be understood in the context of an international forum where most participants recognized the existence of the acid rain problem, yet was totally at odds with the governmental story-line in the domestic debate at that time. Here it is important to know that behind the scenes Dr Martin Holdgate and William Waldegrave, the environment minister at the DoE, sought to use acid rain to introduce what was basically an ecological modernization approach. They most certainly managed to secure some support and even more sympathy, and were also successful in creating several fora where the new ideas could be developed further.[56] Yet the attempt to reconceptualize the general policy approach to environmental issues failed to influence the acid rain decision-making. So Waldegrave and Holdgate could be heard uttering different story-lines. Theoretically there is more to this fact than simply to observe that they clearly addressed different audiences and used story-lines to secure their own goals.

The relative independence of the story-lines from the actors who utter and thereby reproduce these structuring story-lines opens up new ways of understanding the controversy. The story-lines

[55] Sir Hugh Rossi, in *The Times*, 11 Sept. 1986: 9; Dr Martin Holdgate in DoE statement at Munich conference (DoE 1984c).
[56] I will come back to the development of eco-modernist ideas within the DoE in Sect. 4.7.

maintained the structure of the controversy, but the individual freedom to change discursive contributions in actual interaction allowed individual actors to safeguard their own credibility in some situations. It is precisely for this reason that story-lines and specific forms of discursive interaction are better analysed as tied to specific practices, rather than actors. Hence to be able to assess the extent to which the policy discourse of ecological modernization actually affected the regulation of the acid-rain controversy in Britain, we review the story-lines in the context of the evolution of the political process and, more specifically, in the context of the institutional practices within which actors played their argumentative game.

The British acid rain controversy focused on three issues on which actors held competing claims: what damage was done; what role science had to play in the decision-making; and whether the existing regulatory practice was sufficiently equipped to deal with the acid rain issue. Obviously, there are many other practices that have exerted influence on the evolution of air pollution discourse and the way in which acid rain was taken on, but it is impossible to cover the whole domain in similar detail. This study restricts itself to a discussion of practices in these three specific fields. Between them these three issues (the image of damage, the role of science, and the issue of regulation in the strict sense) cover the most essential practices through which acid rain was framed, and illustrate how specific structural discourses (such as science) were drawn upon and discursive strategies were played out.

Finally, before returning to the British case it is important to underscore and slightly elaborate on the immanentist position put forward in Chapter 2. This holds that a discursive order, even if it has solidified in all kinds of institutional arrangements (like laws, organizational routines, or categorizations) requires a constant discursive reproduction to guarantee the continuity of its meaning structures. This implies, first of all, that institutions and argumentative action are to be examined in their interrelations. Practices cannot be analysed as fixed entities. Their meaning is likely to change over time and only by examining the constant debate on the meanings of practices can we establish their role in the overall process of regulation. The immanentist argument also implies that policy change can materialize only if one succeeds in finding ways to overturn routinely reproduced cognitive commitments. In the acid-rain case a particularly important role in this respect was fulfilled by the creation of an image of damage.

4.5. THE CONSTRUCTION OF AN IMAGE OF DAMAGE

> If we had observed these signs in Germany we would have classified them as being caused by air pollution.
>
> German forester, 1984[57]

> We still have not seen the circumstances that we can associate with pollution.
>
> Forestry Commission official, 1988[58]

Acid rain was in at least three respects an anomaly to the prevailing perception of air pollution in Britain: after the 1952 smog air pollution was primarily perceived as a matter of clearing the urban skies; it was related to visible pollution such as smoke and smogs; while the primary aim was to reduce the risks to human health involved. The focus was therefore on reducing the ground level concentrations of potentially harmful substances such as particles, smoke, and heavy metals such as lead. For a long time all important actors (with the notable exception of the Beaver Commission[59]) operated within this 'urban triangle' and consequently

[57] Quoted in Rose and Neville 1985.
[58] A. J. Grayson in HoC Environment Committee (MoE) 1988: 117.
[59] Sir Hugh Beaver chaired the Committee on Air Pollution that was asked to prepare a report on the London smog. The final Beaver Report was issued in November 1954. It now almost reads as a eco-modernist report *avant la lettre*. It asserted that air pollution should no longer be tolerated, that 'enough was known' about its causes and consequences, while technologies for abatement were available. On the basis of an elementary sort of cost–benefit analysis it also purported to show that inaction had great economic costs. Although the Committee saw domestic smoke as the most pressing problem, the Beaver Report also identified sulphur oxides (that later were to play a key role in the acid-rain controversy) as amongst the most harmful components of fuel combustion. It pointed out that, in case of larger power-plants, the sulphur dioxide could be washed from the flue gases and recommended that power-stations should be required to remove SO_2. In the subsequent parliamentary debates it quickly came out within which confines conflicts over air pollution were to be regulated. Interestingly, the government followed most recommendations but did not endorse the general principle of the Beaver Committee, that air pollution should no longer be tolerated. Smoke was to be banned, but on the issue of sulphur dioxide emissions the Clean Air Acts of 1956 straightforwardly dismissed the recommendations of the Beaver Report since they would require 'more than reasonable costs': see Parker 1975; Sanderson 1961.

the emission of SO_2 as such was perceived as a problem only as a ground-level concentration, i.e. when it imposed a direct threat to human health.

For a long time the perspective of the traditional-pragmatist discourse-coalition was really restricted to the confines of the urban triangle. Since the traditional-pragmatist approach had a strongly empiricist inclination, regulatory performance was monitored in terms of the directly observable indicators. The return of the lichens to the city, the clean air, and the increased amount of sunshine, all reinforced the myth of the powerful Clean Air Acts. In the light of this evidence, suggestions of regulatory failure were easily qualified as wild accusations. Given the observable improvement it appeared more sensible to cut expenditure on the constant monitoring of air quality. This was indeed what happened. Between 1977-8 and 1982-3 the money spent on air quality monitoring dropped from £2.59 to £1.3m.[60] Only if there were explicit examples of concrete new incidents should the state step back in.

However, if one were to apply eco-modernist ideas to air pollution one would start from entirely different premisses: the cumulation of gases in the atmosphere had resulted in a high-risk environment. After all, the reports of international organizations like the OECD and the UN-ECE on acid deposition had established that the atmosphere was an unpredictable chemical cocktail of pollutants that was likely to produce all kinds of unexpected reactions. The Government should avoid intervention but should actively promote practices to reduce environmental stress. From this perspective, monitoring did not require additional legitimation but was a basic need. Hence the two different discourses simply framed the very same physical reality in different ways. The traditional-pragmatist discourse stuck to empirical indicators, while eco-modernist discourse was in compliance with the risk society hypothesis that suggested that the new ecological risks typically escaped direct sensual perception.

What clearly inhibited the expansion of the discourse of ecological modernization was the fact that the urban bias had become reified in various institutional practices and concepts. A particularly important practice in this respect was the air pollution monitoring system. Most monitoring stations that were set up under

---

[60] HoL Select Committee on the European Communities 1984: 365.

the 1956 Clean Air Acts were confined to cities: only 150 out of 1,200 were located in rural areas. In 1981 a contented Warren Spring Laboratory, the Government monitoring institute, announced that 'geographical coverage of the UK is being achieved by including some sites from all the conurbations, most of the larger towns and from a sample of the smaller towns, with emphasis on the selection of sites in commercial town centre locations . . . and the denser residential areas.'[61] Even in 1984, Warren Spring argued that it saw its role primarily as keeping a record of SO_2 ground-level concentration patterns in urban areas.[62] This helps to explain the prevalence of the myth that acid rain did not affect the British countryside: for a long time there simply were no relevant data.

As far as SO_2 was concerned the monitoring system at least recognized this substance as a threat. This was certainly not the case for two other acid-rain-related pollutants: ozone and NO_x. In 1984 there was still no picture of the effects and distribution of these pollutants. The institutionalization of the urban triangle and the reification of its bias implied a routinized exclusion that for a long time facilitated a smooth reproduction of the traditional-pragmatist story-line. Hence the power of that discourse had in this respect a very strong institutional basis.

It is not surprising that the correction of the biases in the monitoring network became one of the key demands on the part of the discourse-coalition of ecological modernization. Yet they had the difficult task of proving that air pollution was not an issue of the past and that something as abstract as acid rain also affected those things most dear to the British citizen. It is significant that the first reliable evidence of acid-rain damage was established by the independent monitoring work of the Institute for Terrestrial Ecology (ITE), a prominent supporter of the policy discourse of ecological modernization.[63] At ITE researchers suspected that acid deposition was a more widespread phenomenon and had established their own monitoring network in 1977. Once the—very limited—system of ITE became operative, it showed that the acidity of rainfall in northern Britain was comparable with that recorded in southern Scandinavia. Scientists from the ITE and the

---

[61] Quoted in Weidner and Knoepfel, 1985: 262.

[62] HoC Environment Committee (MoE) 1984: 138.

[63] Their evidence was first presented in Nicholson *et al.* 1980 and later in the influential article by Fowler *et al.* 1982.

NERC (but later also the Royal Commission, the HoC Environment Committee, and FoE), complained about the lack of rural data and data on NO_x and ozone.[64] They realized that better domestic monitoring was essential to break away from the traditional-pragmatist and *post hoc* approach. As a result the Government promised to extend its SO_2-monitoring programme and improve the monitoring of other pollutants such as NO_x and ozone following recommendations by the UK Review Group on Acid Deposition, the Royal Commission, and the HoC Environment Committee in 1984.[65] It was the first evidence of a process of restructuring of air pollution discourse.

Explaining the Persistence of the Urban Triangle

The explanation for the persistence of the urban triangle cannot exclude the level of strategic discursive action. The CEGB used the lack of data to reinforce its argument that acid rain was an ill-considered problem. In its evidence to the HoC Environment Committee the CEGB states that 'recently there were 104 occasions when a particular value (for ozone pollution) was exceeded in Germany and only once in the United Kingdom'.[66] This fact is hardly surprising given the lack of any systematic measurement.[67] The government's response to acid rain likewise was not only a result of the institutionalized biases. It was also based on the active reproduction of the well-established myth of the proud record

[64] NERC in HoC Environment Committee (MoE) 1984: 222. Royal Commission 1984: 116.

[65] DoE 1984b: 18; DoE 1984d: 17. This change of heart was not entirely voluntary since the Government had to comply with new rules about air-pollution monitoring developed by the EEC. In 1990 the British monitoring network included 295 sites for smoke and SO_2, 11 sites for NO_x, 17 for ozone, and another 32 specifically for monitoring 'acid deposition' (DoE 1989: 12).

[66] HoC Environment Committee (MoE) 1984: 18.

[67] The lack of monitoring was not only the result of the perhaps politically induced bias towards the urban realm. The research institutes that needed the data for their work and could have broken through this bias did not like to debit their research accounts for the monotonous acquisition of data (see e.g. the evidence presented by the NERC to the HoC Environment Committee (MoE) 1984: 265. The high costs of making longitudinal records available for scientists would therefore have to come from some other source. Yet, in the absence of an understanding of the problem, who was going to take the initiative if not the scientists who suspected the possible damage?

of the British pragmatist response.[68] This myth has a remarkable resilience and can be found in air pollution discourse to this day.

In their comprehensive comparative study of air-pollution policy in Western Europe, Knoepfel and Weidner conclude that the SO_2 pollution levels in Britain are comparatively high and that the number of casualties due to SO_2 pollution is still much higher in Britain than elsewhere.[69] There have been relative improvements in British air quality, but in international perspective the endless reiteration of the proud record in the context of the acid rain problem is best seen as a combination of a routinized form of historical comparison (instead of international comparison) combined with an active playing down of a new problem using a metaphorical form by others. The point is, however, that the one cannot be separated from the other: if you think you have made great improvements new problems may indeed appear relatively minor. This is what we see in the traditional-pragmatist story-line on acid rain. It positioned the new issue as analogous to the urban air pollution of the 1950s (which killed 4,000 people). In this perspective the reaction to the London smog is subsequently put forward as a set-piece of effective pragmatist regulation, the implication being that the same institutions could be trusted to find the appropriate answer for the alleged acid-rain problem too. Acid rain was thus denied any special status despite the fact that both the 1976 statement and the OECD report of 1977 had revealed that air pollution had taken on new characteristics and should be understood on a broader geographical scale.[70] The effect of the reiteration by the government of the proud record is to be seen as a ritualization of the problem, in the sense in which Edelman used the term, whereby political conflicts are stabilized and new sources of escalation are effectively suppressed.[71] The urban metaphor had clearly become a political tool. It was a trope that was structured in public consciousness thanks to the most elementary rhetorical technique: repetition. The proud record thus effectively became a cliché which implied that actors would no longer consciously consider the actual meaning of what was being said.

Potentially acid rain stood to illuminate the structural deficiencies

[68] For the empirical evidence of the usage of the proud record, see Sect. 4.3.
[69] Knoepfel and Weidner 1985: 38–9.
[70] Cf. DoE 1976.   [71] Cf. Edelman 1971: 21 ff., 142 ff.

of the traditional-pragmatist regulatory regime. It was clear that more was needed than the mere presentation in the political arena of some abstract acidity statistics showing increases in deposition levels. In the face of a dominant policy discourse that sought to understand new problems in old categories, the task of the governmental critics was to change the very cognitive frames within which problems were perceived. The WCS might have provided part of the intellectual argument, yet to be persuasive one needed the support of empirical data as well. Hence a key moment in the struggle for discursive hegemony became the creation of a persuasive alternative image of damage, images that would have enough appeal to break through the deeply embedded and comfortable idea that air pollution was an issue on its way out. An image was needed that would break the reliance on direct observation and would elucidate the rather intellectually demanding position that air pollution was now a matter of many different substances causing diverse and long-term problems at various (and often far away) places. It was in this context that the issue of tree health played a decisive role. Despite the traditional-pragmatist claim that acid rain was an issue about fish dying in Scandinavia, over the years dying trees came to dominate the public image of acid rain damage.

Reports in newspapers and general science magazines like *New Scientist* most certainly contributed considerably to the creation of a new image of damage. More concrete political reflection was brought about by the visits of MPs to forests in Scandinavia and Germany.[72] As a result of these reports and visits the agenda started to change. The question became whether the damage observed abroad also occurred in Britain. Here the critics of Government not only had the burden of proof, they also had to fight some surprisingly persistent common-sense ideas suggesting that the 'prevailing western winds' blew all pollutants away. This idea in fact still dominates public discourse to a considerable extent. Yet in the policy debate the image of tree health changed considerably during the 1980s. Tree health is one of the issues where eco-modernist concepts emerged relatively quickly and it is therefore relevant to examine the issue of tree health in some detail.

---

[72] See the frequent references of MPs to these visits both in the Fourth Report and the minutes of the discussions (HoC Environment Committee 1984).

The State of British Trees

The issue of tree health was first raised by Friends of the Earth Scotland in May 1984. They had brought a West German forester to examine British trees. He argued that 'if we had observed these signs in Germany we would have classified them as being caused by air pollution.'[73] In 1985 and 1988 FoE organized its own Tree Dieback surveys that indicated that the symptoms of decay could also be found on British trees.[74] However, these surveys could easily be disqualified as being partisan and unscientific. The only institute that monitored tree health on a regular and professional basis was the Forestry Commission. The Forestry Commission (FC) argued that it had not 'found it necessary to invoke the acid deposition hypothesis' in explaining damage and that its surveys 'showed no symptoms similar to those seen in central Europe.'[75] It was this professional assessment that for a long time formed the backbone of the Government's claim that acid rain did not affect Britain.

The HoC Environment Committee, concerned after having seen the scale of forest damage in Germany, was not satisfied with this assessment. It recommended that the Forestry Commission conduct a survey on the same lines as on the continent.[76] The FC was clearly not at ease with its new role, as the following quotation from a generally businesslike account indicates: 'The determination of certain observers to prove that pollution effects exist where careful observation shows them not to, has been an interesting feature of the decade'.[77] Even with the new indicators of crown density and needle yellowing, the surveys by the Forestry Commission at first did not indicate any symptoms similar to those seen in central Europe. This quickly changed. In 1985 still 'no objective evidence of damage caused by other than local sources' was found whilst it was also stated that 'general surveys of the condition of

---

[73] Quoted in Rose and Neville 1985.   [74] Ibid.; FoE 1988.
[75] HoC Environment Committee (MoE) 1984: 206, Forestry Commission 1985: 24; see also HoC Environment Committee (MoE) 1984: 201.
[76] HoC Environment Committee 1984: xl, xli. This call came rather late in the day, since the Forestry Commission had just had to remodel its survey methods following the protocol of the UN-ECE working group on the effects of air pollution on forests. The protocol was compulsory for all EEC countries.
[77] Forestry Commission 1985: 1.

tree crowns appear likely to be of little use.'[78] Yet the 1986 Report in fact asserted that 'It appears that there has been a marked decrease in that assessed crown density since 1985 and that the position is also substantially worse than in 1984.'[79] The condition of trees could now only be classified as moderate.[80] By 1988 the change in the Forestry Commission's assessment was complete. The annual survey showed that 'the overall figures are now similar to those reported from central European countries.'[81] Hence between 1985 and 1988 the Forestry Commission had completely changed its assessment of British tree health, as is illustrated in Fig. 4.1. It is unlikely that this graph simply reflects a revolutionary change in the state of the British trees. The graph in fact indicates some sub-political processes that greatly influenced the acid rain controversy.

Tree health surveys are a nice example of sub-politics, since the new outcomes and especially this graphical representation, had profound political implications. It radically changed the input of the Forestry Commission to the political process and disorganized the government's rhetoric of factuality. In 1984 the Government could still fend off its critics by arguing that the experts at the Forestry Commission had not found any signs of damage. In 1988 new publications of findings suggested a serious decline in the condition of British trees. This entirely changed the image of air pollution damage, opened up the problem of its definition, and thus contributed to a new, and more reflexive understanding of the acid rain problem. No longer could it be maintained that acid rain was a Swedish problem, and the lack of understanding clearly called for research on pollutants other than sulphur oxides.

The development sketched in Fig. 4.1 does not simply reflect objective changes in the condition of British trees but also reflects a conceptual and cultural shift that took place within the Forestry Commission. It should be remembered that the British Forestry Commission is not to be confused with institutes like the Nature Conservancy Council that are primarily interested in conservation issues. The Forestry Commission was founded in 1919 following the shortage of pit props during the Great War. The statutory goal of the Forestry Commission was therefore simply to grow timber

---

[78] Forestry Commission 1986: 3.   [79] Forestry Commission 1987: 31.
[80] Innes *et al.* 1986.   [81] Forestry Commission 1988: 28.

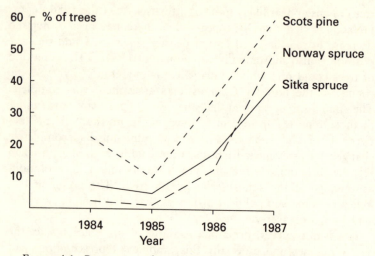

FIGURE 4.1. *Percentage of trees with more than 25% reduction in crown density over four years of tree health surveys.*

as a crop. The Forestry Commission produced softwood (timber) primarily on upland plantations, cultivating trees like conifers. Of course, it had always monitored tree health but only to be able to make growth curves: it needed to know how to grow trees quickly and it needed to know at what age they were likely to start to show signs of ill health so that they could be felled beforehand. Traditionally, the Forestry Commission's contribution to research was therefore dominated by physiological and pathological research for mechanisms that could explain why a tree did not grow. All that came out of this reductionist type of research was evidence that related tree health to drought (water deficit).

Although it is fair to say that during the 1980s the FC became more active on environmental issues, it is also evident that the FC was very suspicious of the sudden public attention given to tree health. The FC derived the logic of its work from its own storyline of growing timber as a crop and was therefore reluctant to start conducting surveys examining crown density and needle yellowing. The older Forestry Commission officials, who were used to the relative quiet, were highly uncomfortable with the public attention. The scaremongering of groups like FoE had the effect that they denied the subject-positioning in the discourse of air pollution and put an extra weight on their own professional ethic of sticking to the facts. If there was one thing they wanted to avoid, it was being associated with pressure groups that 'did not know about trees.'

In this respect the acid rain controversy was a cultural conflict. Friends of the Earth introduced young and eloquent urbanites who explicitly politicized the issue of tree health and accused the FC of a lack of responsiveness to issues of natural degradation. The scientific basis for these accusations was in fact rather dubious. Having exchanged ideas about what a healthy tree looks like at the international NGO conference at Eerbeek (the Netherlands) in May 1985, FoE decided to do its own Tree Dieback Survey. The FoE survey was published in October 1985. FoE were quick to point out that the evidence was only circumstantial, that it had used a foreign methodology, and that the survey had been conducted by volunteers. Still, the survey results were alarming enough with 69 per cent of beech and 78 per cent of yew showing some sign of ill health.

FoE intended to arouse a public outcry, or, to put it differently, had aimed to provide the abstract acid-rain issue with concrete images of damage. That was also why Chris Rose, air pollution campaigner at FoE, decided to do a survey among British broad-leaf trees. He realized that the imported conifer trees at the upland plantations were not particularly popular in Britain:[82] 'We needed to find something which really affected the core values of the British establishment . . . That is partly why we went for trees. It was also a problem that affected the South East of England whereas acidification of freshwaters was mainly to do with Wales and Scotland.'[83] The effectiveness of this intervention was most certainly reinforced by the fact that William Waldegrave likewise hoped to use the image of the dying tree as a lever in bringing about a change within the government. Within a week of the FoE results becoming public, Waldegrave announced plans for emergency tree surveys. In November 1985 Waldegrave set up a UK Tree Health Group, with representatives from the NCC, FoE, NERC, and FC. It quickly agreed on changes in the survey methods and effectively broke the expert monopoly of the FC.

[82] This trick was repeated in 1988 with the Windsor Great Park Beech Health Survey (FoE 1988).
[83] Mr C. Rose, interview.

Waldegrave's intensification of British research was not meant as a fact-finding mission but was a clear example of using the aura of scientific activity as a means of persuasion. As he stated: 'One needed to have some good people here working on these things . . . Politicians, decision makers would be much impressed by what they would see if they would go to Lancaster University laboratories and see stunted trees . . . You know, they would believe it.'[84] The FoE tree health survey helped to create a new image of damage and was arguably the most important intervention by FoE in the acid rain controversy. Now there were two rival views on something as fundamental as what a healthy tree looked like in Britain. The seemingly innocent and undoubtedly amateurish FoE survey thus opened up debate on the assumptions that underpinned the FC's view that British trees were healthy. The FC now had to explicate its assessment, convince others rather than simply impose its expert judgement on policy-makers, in short it had to become reflexive.

Under pressure from Whitehall the Forestry Commission had to adjust its position. The FC had always argued that Britain was a maritime country and was as such not comparable with the Black Forest but with Norway. This implied that continental criteria for forest health were not necessarily valid, which is certainly a legitimate argument. FoE on the other hand, followed their continental advisors and were trained to look at crown density as an indicator of tree health. A thin crown simply was attributed to acid rain; a quite unacceptable assumption for the Forestry Commission.

Now the irony of this example is that although the officials of the Forestry Commission were certainly right to argue that British trees cannot be compared with trees in continental Europe they were politically forced to look into acid rain related damage. Yet once the Forestry Commission was forced to work on pollution surveys it increasingly became aware that something strange was going on. The Forestry Commission started to notice crown thinning too, but how should it be interpreted? How would you know whether the British trees were in decline if you had no historical records? The validity of the image of a healthy tree from the Swiss *Sanasilva Guide* was doubted because of geographical differences.[85]

---

[84] Rt. Hon. William Waldegrave, interview.
[85] Dr J. L. Innes, HoC Environment Committee (MoE) 1988: 112.

In fact only long-term data could provide an idea of what a healthy spruce looks like. Yet in the face of the public attention to tree health, results were required immediately. Arguing that a responsible assessment could not be made in the short term would not do: at some point an institute can lose its credibility by not making a statement, even if it is convinced that its own scientific standards would require a more reticent reading of the findings. Here the public concern clearly forced the FC to change its stand.

This cultural disruption of the institutional routines had several aspects that should not be reduced to one another: personal changes, an opening up of institutional practices, and, last but not least, a new conceptual frame. In 1987 the Forestry Commission had started to work with a multiple stress hypothesis. It argued that

No single mechanism is seen as responsible for the tree decline, rather the decline is the result of the cumulative effects of a number of stresses. There is a growing amount of evidence that some factors predispose a tree to damage [i.e. climatic change] whereas others incite or contribute to the damage [i.e. drought, frost, pollution].[86]

This epistemological shift allowed in pollution as a serious variable and was as such part of the much wider cultural shift within the Forestry Commission. I would argue that it not only changed the interpretation of facts but also affected the very creation of facts. The multiple stress hypothesis does not necessarily imply the acceptance of the thesis that pollution is responsible for tree dieback. It merely leaves open the possibility that pollution is implicated. This changed the practice of observation and assessment. In the years before, the examination of crown density and needle yellowing, indeed the whole practice of tree health surveys, had been associated with pollution and the acid deposition hypothesis. In that context to notice crown thinning would support the scaremonger and went against the traditional forester's interpretation of ill-health as related to natural factors like drought or frost. The fact that the multiple stress hypothesis separated observation from evaluation might well have changed the individual attitude towards surveying and thus helps to explain the upswing evident in Fig. 4.1.[87]

---

[86] Innes 1987: 25.

[87] It is important in this respect to emphasize that both Dr Freer-Smith and his Dutch counterpart Dr den Boer stressed the huge impact of surveyors' training on the actual survey results.

The acceptance of the multiple stress hypothesis in the circles of the FC is a seemingly small fact but proves how eco-modernist story-lines penetrated the British FC. It is the acceptance of these seemingly technical notions that indicates the penetration of ecological modernization in the realms of British pollution politics. It indicates the institutionalization of a discourse, shows how this changes the self-understanding of the Forestry Commission, and makes it less likely that practices can simply be reversed to operate in the traditional-pragmatist mode. Although in terms of environmental policy-making these changes in the cognitive basis of judgement were the more fundamental, they received far less attention from the press and academics than the new outcomes of the monitoring activities that were still stuck in a naturalistic construction of the problem.

4.6. THE BRITISH SCIENCE–POLITICS INTERFACE: THE VEIL OF SCIENCE FOR PROOF

I don't think we expected SWAP to make great strides, new steps, in understanding and so on.

Dr G. Howells, CEGB scientist

Had SWAP not happened, then the abatement of 1986 would not have occurred.

Professor F. T. Last, ITE

Environmental issues like acid rain conceal complicated scientific questions. Decision-makers therefore often commission large research projects to get answers to questions concerning the precise nature of the physical problem. By so doing they seek to create an authoritative basis for their own regulatory response. As the account on tree health already indicates, science thus essentially delimits the realms of political decision-making. The science–politics interface, as it is sometimes called,[88] is therefore a particularly important field of practices through which discursive power is exercised and which should accordingly be included in a discourse analysis of the political process.

This task can be taken on in different ways. The most

---

[88] See Boehmer-Christiansen and Skea 1991: 47.

straightforward way is by focusing on the way in which policy-makers employ scientific knowledge. In this manner Boehmer-Christiansen has argued that science was used as a fig-leaf for policy in the context of the British acid rain controversy.[89] Yet this is, of course, precisely the type of analysis of political power that Foucault criticized. Implicit in the fig-leaf metaphor is the idea of an almighty state (Foucault's sovereign) that uses science to its benefit. Power is thus seen as an essentially centralized entity. What is more, it is based on a purely uncritical understanding of science. A relational analysis of the science–politics interface suggests that power should be analysed as inherent in the knowledge claims and the various practices through which specific scientific claims gain authority and credibility.

In a more recent publication by Boehmer-Christiansen and Skea the relationship between science and politics receives a more subtle treatment.[90] Here they emphasize the centrality of the question of the communication of scientific knowledge. Boehmer-Christiansen and Skea show how this communicative process can be seen as a multi-layered process of the condensing and sifting of information, in which specialized scientific knowledge is processed through all kinds of intermediate links, such as review papers, executive summaries, or newspaper articles. Here power is no longer a function of the sovereign but is seen, if you like, as inherent in all kinds of social practices. However, what remains is a rather classic conception of science and scientific knowledge. The emphasis is on the way in which basically unquestioned scientific insights are (mis)represented. As Wynne has often observed, science is itself 'more subtly malleable than [is often] suggested' adding that 'there is no preordained and objective definition of the boundary between science and politics.'[91] Communication of knowledge is obviously not a one-way affair but is essentially an interactive process. In straightforward institutional terms one may note, for example, that much research in this area is commissioned by policy-making institutes, which implies that scientists work within frameworks that were defined by decision-makers. Furthermore, the actual research is often closely monitored by a steering group that, in various ways, exerts influence on the actual production of

---

[89] Boehmer-Christiansen 1988*a*.
[90] Boehmer-Christiansen and Skea 1991: 47 ff.   [91] Wynne 1992*a*: 746.

scientific knowledge. But even if policy-makers do not actively interfere one may expect a certain degree of anticipation.[92]

The analysis of discourse-coalitions suggests the need for an examination of the way in which credibility is allocated in science and in policy and the way in which the various actors active in these domains influence these definitions. The assumption behind the discourse-analytical approach to the science–politics interface is that these coalitions and their story-lines are what provide the essential context which allows actors (scientists and policy-makers alike) to keep up and reinforce certain beliefs or trust in certain concepts or approaches. The science–politics interface is then analysed as a set of discursive practices in which rival groups of actors seek to impose specific authority claims and understandings of the problem on a policy community. Here one could draw on an important distinction made by Jasanoff.[93] She suggests distinguishing 'research science' from 'regulatory science'. The interesting idea behind this distinction is that in order to understand the science–politics interface we need to break away from the classical division of 'proper science' and 'science for policy'. Implicit in this model is the idea of a singular model of appropriate scientific practice that would function under normal circumstances. Research in the sociology of science has given ample evidence to suggest that such natural conditions do not exist and have never existed. Indeed, the work of Merton and Fleck indicated a long time ago that social inspiration is essential to the development of scientific knowledge.[94] Jasanoff comes up with a distinction that more explicitly takes the sociology of scientific practice as a starting-point. Research science is conducted in established, well-defined, and socially stable networks that have a strong set of integrative practices such as peer review, and is held together by shared epistemological ideas, and by cognitive and practical commitments. Regulatory science on the other hand, because of the nature of the questions it seeks to answer, is characteristically conducted in networks with a less well-defined set of operational practices or

---

[92] The interactive nature of the process of communicating knowledge also suggests that one should not use the words knowledge mediation or speak of the translation of scientific knowledge, both of which seem to miss the interactive nature of the process in which scientific knowledge is taken up in the political process and policy-making concerns affect the science.
[93] See Jasanoff 1990.   [94] Merton 1973; Fleck 1979.

epistemic commitments. It would be a mistake to see this distinction as necessarily leading research into pure relativist alleys. The identity of research-science communities is often the materialization of reasoned choices and reflection upon the faults and mistakes of previous experiences. At the same time the distinction avoids the pitfall of conceptualizing science and politics according to the well-established hierarchization of pollution science as subordinate to proper science. It would be more useful to see the two as distinct practices that each have to meet their own functional requirements and can be more, or less, established, and can have a varying prestige or credibility.

Finally, to facilitate the assessment of the reflexivity of the science–politics interface in the acid rain controversy, it is useful to recall another rather simple distinction between two types of legitimation of scientific work. In an article on the role of science in policy-making Clark and Majone distinguish 'numinous' and 'civil' legitimacy of scientific claims.[95] The numinous form of legitimacy is derived from 'superior authority held to be beyond questioning by those who endure the consequent exercise of power.' Civil legitimacy, on the other hand, is determined by the degree to which scientific claims secure credibility in public debates and is related to the way in which they are played out against other expert opinions or general political or social concerns.

The British Ideal of a Science-based Policy Approach

According to what I call the British Science-based Policy Approach (SPA) air pollution regulation should not only be technically effective and economically feasible; above all, policy interventions have to be legitimized by scientific evidence. These three requirements are in fact strongly interrelated. Science has to come up with the evidence for the existence of an environmental problem to rule out unnecessary expenditure. The SPA to environmental politics was clarified by the DoE in Pollution Paper No. 11:

People are naturally very much concerned about the effects of pollution on health . . . It is inherent in our society that such pressures should arise, but to accede to them unquestioningly could often involve a waste of resources as well as the possible loss of activities and benefits on which

[95] Clark and Majone 1985: 16.

society places value . . . it is important to ensure that the standards being imposed do not rest on an unsound scientific justification or require disproportionate economic costs, since this would make it difficult later to introduce further measures, however well founded. In explaining standards however, the risk of gross misinterpretation of data and the need to avoid disclosure of truly confidential information need to be borne in mind.[96]

The dominant role for science was legitimized as a way of keeping pollution control out of the sphere of corporatism and pressure-group activity, but in fact it was much more deeply embedded in a set of institutional practices. Over the years both Conservative and Labour governments have reinforced this practice. However, British ecological modernization took up the issue of the SPA. In 1984 the Royal Commission argued:

Environmental regulation is . . . not simply an issue of scientific facts, and it is a delusion to imagine that decisions, even in an ideal world, could ever be based on complete impartiality and objectivity. We expect government to behave like a manager, not a scientific referee . . . Rigorous scientific 'proof' should . . . not be regarded as a sine qua non.[97]

And further on:

Faced with inconclusive evidence, the scientist is trained to respond with caution. However, evidence that is not conclusive when judged by the conventions adopted in scientific research may yet be a reasonable cause for concern to those who have to act on it outside the laboratory. The politician or manager who must decide what action to take now cannot wait for the rigorous proof that is properly demanded by a referee for a scientific journal.[98]

In other words, the Royal Commission was perceptive to the particularities of pollution science, warned against an easy conflation of the discourse and practices of pollution science with established research-scientific traditions and called for the recognition of the independent requirements of this field. The way in which the SPA was employed in the acid rain controversy strongly suggests that the SPA required a particular sort of science that, following its own discursive logic, had an affinity with the arguments of some pressure-groups whilst rebutting the arguments of others.

---

[96] DoE 1977: 7–8.
[97] Royal Commission 1984: 22.   [98] Ibid. 176.

In this sense the SPA is closely linked to the Government's insistence on confidentiality, again something which was fiercely contested by eco-modernists. The commitment to confidentiality effectively excluded the general public but also possible counter-experts and thus inhibited a public debate on the science of environmental issues. The practice of secrecy was frequently challenged, most notably by the Royal Commission. In its Tenth Report the Royal Commission basically made a case that science now required civil legitimacy arguing:

... we have seen much evidence of a growing professionalism in many voluntary organisations in the environmental field and with it an ability to evaluate and present technical and scientific reports which compare favourably with those of officially sponsored researchers. In our view, to deny access to data on grounds that 'the public' is not competent to make 'correct' use of them is neither a tenable nor an acceptable position.[99]

In the Government's reply to the Tenth Report the Government met its critics half way, maintaining that

Accurate information about the state of our environment, and reliable scientific analysis of the way in which environmental systems work, is essential for informed judgement. Risks should be measured as objectively as possible—with uncertainties made equally explicit. But the Government also accepts that people's feelings may not coincide with the scientific estimate of the risk. Environmental policy must take public feelings into account. Where they are the result of misconception, only information—which is accepted as unbiased and derived from open processes—is likely to remove them. Where they are the considered results of deeply-held values, they have to be taken very seriously ...[100]

The distinction between the facts and deeply-held values is, of course, a highly problematic one[101] yet here it is the definition of the SPA practice that concerns us. This remains basically indeterminate and leaves open which importance should be attached to scientific and non-scientific perceptions of the problems at hand. Hence to assess the effects of the SPA practice on the regulation of the acid-rain problem we have to inquire how the SPA practice was actually taken up in the political process.

---

[99] Ibid. 27.   [100] DoE 1984b: vii.
[101] See Ch. 2. For a critique of the fact–value dichotomy see Fischer 1980; Bernstein 1976: 24 ff; see also Wynne 1982; Schwarz and Thompson 1990.

The SWAP Project

The specific functioning of the SPA practice in the context of the acid-rain controversy will become apparent if we re-examine the 1986 decision to retrofit three power-stations with FGD. This decision is usually seen as a breaking-point in the controversy. Both the CEGB and the government argued that this decision was based on the emergence of new scientific evidence that had convinced them that the installation on SO_2 scrubbers would indeed have a positive effect on the acidified Scandinavian environment. One could see this reference to new scientific evidence as actively employed rhetorics or see it as a case of using science as a fig-leaf. Yet there is in fact every reason to see this as an example of how power is routinized in the cognitive commitments and functional rules of practices of a given domain.

A proper understanding of the 1986 decision has to go back to at least September 1983. In that month a major conference on acid rain was held at the Royal Society. The conference had been called by, amongst others, the Chief Scientist of the DoE, Dr Martin Holdgate. It discussed the state of knowledge of the effects of sulphur and nitrogen compounds on the environment. Although the initiators had meant the conference to create a strong basis for action,[102] it ended up criticizing the partial and imprecise data available. As such it provided the perfect forum for Lord Marshall (then Sir Walter), chairman of the CEGB and himself a fellow of the Royal Society, to announce the launch of the SWAP research project. The research on acid rain, it was announced, was now given into the hands of the 'most prestigious scientific institution of the world', which would most certainly produce the data that the experts so dearly missed.[103] The SWAP research effectively put a moratorium on any decision-making. It also reinforced the SPA practice since Lord Marshall announced that, whatever the outcome of this research would be, the CEGB would accept it.

It has been observed that the Royal Society (and the Scandinavian academies of science) thus essentially became the arbitrator in the dispute. However, what is often missed is that the launch of SWAP meant that the CEGB effectively imposed a specific set of research-

[102] Dr M. Holdgate, interview, see Beament *et al*. 1984.
[103] Cf. Anon. 1983*b*: 85.

scientific criteria on the decision-making of a policy issue. For instance, the introduction of SWAP pushed aside the previously expressed need to take account of deeply held beliefs and reconfirmed the numinous legitimacy of science instead.[104] SWAP effectively precluded any application of the eco-modernist ideas that were reluctantly unfolding in the environmental domain, let alone admitting the claim that pollution matters were issues that called for social negotiation. As the SWAP research became the linchpin, the traditional SPA practice that seemed to be up for reorganization recovered lost ground. Indeed it now received the support of the ethos of the Royal Society, a highly respected British institution.

Yet the disciplinary effects of the inauguration of SWAP not only implied the exclusion of non-scientists or moral criteria from the debate on acid rain. The introduction of the Royal Society into the heart of the British science–politics interface also inhibited a meaningful discussion of eco-modernist concepts like the multiple stress hypothesis. Multiple stress was not an environmentalist invention. It was a concept that was in good currency both among biologists and especially among scientists active on pollution issues. Professor Last (ITE) emphasized that one of the problems with pollution-related science is 'that you by and large have to work on the strength of circumstantial evidence: it is difficult to get absolute proof. It really is about the threshold of circumstantial evidence that you accept.'[105] In itself it could have been an appropriate basis for a revised SPA practice. Holdgate maintained that 'It is the question of the probability that the system falls apart which is important. Not the likelihood that one factor causes another.'[106] However, the launch of SWAP had almost the opposite effect. While SWAP was in progress the SPA came to emphasize the need to come to a precise understanding of the causal mechanisms that were implicated in the acid-rain phenomenon. At the launch of SWAP Lord Marshall actually argued: 'I think it is important as a matter of principle to do everything we can to

---

[104] Seen from this perspective one may wonder whether it was, from an eco-modernist point of view, a good decision to take the acid-rain issue to the Royal Society in the first place: see Marshall 1983.

[105] This position was reconfirmed in the interviews with Pearce, Southwood, and Freer-Smith and is also evident in the statements by NERC scientists at HoC Environment Committee (MoE) 1984: 228.

[106] Dr M. Holdgate, interview.

bring our understanding of non-nuclear environmental matters up to the same standards that we have of nuclear matters.'[107]

The introduction of the prestigious Royal Society into the sphere of pollution politics was in effect a particularly powerful discursive interpellation on the part of the CEGB chairman. To be recognized as evidence by SWAP, research had to fulfil standards that were most certainly not the standards that the scientists of the mundane world of pollution science employed. At a time when many biologists thought that the available evidence legitimated action, and nearly all scientists who had been involved with pollutants would say that evidence was sufficient and disagreed with the government's approach, SWAP introduced the standards that had their roots firmly in the experimental science that had indeed given the Royal Society its world-wide fame.[108] Scientists may of course argue that 'environmentalists may well be right, ultimately, but their evidence is scientifically unsound',[109] but the eco-modernists had a compelling case when they argued that political decision-making should find its own way to break through the scientist's concern for detail. Indeed, Dr Gwynneth Howells, one of the most outspoken scientists working at the CEGB research laboratories at Leatherhead, replied to the question asking when she thought they would know enough about the acidification problem: 'We will never know enough.'[110] Yet significantly, she added that action should be taken if there was a willingness to pay. In other words, she subscribed to the civil legitimacy requirement of science and implicitly called upon the government to act as manager.

The initiation of SWAP determined the discursive confines. Under the veil of the need for proper science and exact understanding the criteria of laboratory science were imported into the realm of pollution science and policy-making. On the basis of my research it is difficult to say whether SWAP was a matter of reinforcing existing rules of what constituted proof or whether it actually introduced the tough standards into the domain of pollution science. What can safely be maintained, though, is that some politicians must have experienced a discursive affinity with those scientists who argued that environmentalists were rushing to conclusions (similarly, those politicians who were concerned about

---

[107] CEGB 1983.
[108] This came out of the HoC Environment Committee hearings of Spring 1984.
[109] Dr R. Pearce, interview. [110] Dr G. Howells, interview.

the state of the natural environment must have felt a discursive affinity with scientists who argued that enough was known). In other words, the centrality of SWAP should be understood in the context of the broader traditional-pragmatist discourse-coalition that included the dominant political forces of that day.

In Chapter 2 I argued that structural support and discursive coherence cannot simply be assumed but are in need of constant reproduction and constantly have to be defended against rival claims. Close examination of the acid-rain controversy hints that even within this structural asymmetry in power relations, eco-modernists might have made a difference. A particularly significant moment in this respect was the meeting at Chequers in May 1984. It was the kind of meeting at which the defining characteristics of SPA might have been altered. In the event the Prime Minister, Margaret Thatcher, invited leading scientists to Chequers to get an update on the state of knowledge on acid rain. As it happened this meeting further reinforced the commitment to SWAP. Yet one wonders whether different discursive strategies might have changed the course of events. The major research institutions working on the topic were represented. Dr Martin Holdgate, Chief Scientist at the DoE, who was personally convinced that immediate action should be taken, presented a paper trying to prove the relationship between SO_2 emissions and forest dieback in the Black Forest. He found his arguments countered by CEGB representatives who pointed to many contradictions in the phenomenon of forest dieback. This apparently put off the Prime Minister. The second actor who could have conveyed the ecological modernization perspective was the NERC representative Dr Hermann Bondi, the science director of NERC but not himself active on acid-rain science. His presentation was perceived by respondents to be very weak.[111] The CEGB, on the other hand, presented its view strongly, arguing that the UK was probably guilty of killing fish in Scandinavia but that evidence was not conclusive. It emphasized the lack of knowledge and referred to the SWAP project of the Royal Society which was investigating the issue. According to informants the Prime Minister concluded the meeting saying that she now knew that there was no firm evidence

---

[111] Interestingly Professor Last, himself active on acid-rain-related research at ITE and a very outspoken scientist, was originally approached to represent the NERC but at the last moment he was replaced by Dr Bondi.

and that therefore the critics should stop harassing the government.[112] Waldegrave, who had hoped that the Chequers meeting would change the disinterestedness of the Prime Minister on the issue, argued 'That Chequers meeting was something of a disaster. . . . It may be that the presentations were not good enough. I think it is possible to present that science in a more formidable way than we did at that occasion.'[113] In fact Waldegrave and his advisors faced a discursive dilemma at Chequers. They could either argue their case within the prescribed confines of scientific discourse or they could try to contextualize this discourse, question the numinous legitimacy of science and explicitly bring in policy choices. In that case eco-modernists could have argued their case on their own terms. The most significant questions would then be what would be the appropriate epistemological basis for SPA and how far one should take a precautionary stand in dealing with new environmental issues like acid rain. In actual fact one could have drawn different conclusions from the meeting if the precautionary case had been on the table. The fact that the CEGB itself admitted that it was probably guilty could then have been a basis for action. This would of course have shifted the debate away from the science and would have brought the more action-oriented policy discourse of ecological modernization into the centre of debate.

The boundaries of science and policy are by no means clear but constantly have to be renegotiated. In this case the employment of scientific discourse for managerial purposes stands out. By drawing on his credibility as scientist Marshall the Manager intermingled with Marshall the Scientist. Similarly, Thatcher the Prime Minister made up her mind drawing on Thatcher the Science Graduate. The assumption under the central role of science was that this is a universal language that can be meaningfully employed by actors with some scientific background. Yet this is a highly questionable postulate. For instance, Dr Howells reported that Dr Peter Chester, her own director of Research at the CEGB, simply could not believe that CEGB emissions could be responsible for dead fish. She continued: 'Sometimes it is hard for a physicist to understand the complexities of biological systems where you get secondary, tertiary or quaternary responses. You know,

[112] Lord Marshall, interview.
[113] Rt. Hon. William Waldegrave, interview.

you have got a whole lot of intervening things in the middle.'[114] Southwood on his part argued that 'You should realise that with these kind of complex ecological situations it is not a matter of sudden realisations, as with laboratory experiments . . . you have to accumulate more and more things until you say "Yes, evidence is now overwhelming".'[115]

More evidence for the essential assumption that science is a sort of Esperanto that uses symbols and epistemological rules that are the same wherever they are applied, comes from Lord Marshall's suggestion that environmental science was basically an immature field of study that should be brought up to the standards of nuclear physics. Yet persuasion at meetings like Chequers is not a matter of logic but at least as much of authority and presentation. The employment of science as the solely credible discourse at Chequers was a highly ritualistic affair that served to reproduce existing power relations. Obviously, this needed to be reconfirmed at that specific moment and most certainly the poor performance of potential eco-modernist authors did not help. Yet here the challengers tried to express an eco-modernist interpretation but forgot to insist on their freedom to choose the language in which to do so.

The suggestion that the role of science should be understood as a ritualized practice should be sharply distinguished from the assertion that science was used as a fig-leaf. I would argue that the employment of science is to be understood as a core element of the discursive style of British pollution politics. It is not something that is particular to the acid-rain case, it has a long history. Yet the portrayal of SPA as ritual is nevertheless a demystification of the SPA as projected by the British Government. It seems significant in this respect that when asked when they were persuaded, nearly all respondents emphasized the key role of visits to specific research projects. Meeting the scientists and getting a visual representation of their work seems a necessary step in the process of cognitive change. Hence Waldegrave argued 'It may strike you as comic but it was a very high priority to me to get the Prime Minister to visit Scandinavia . . . The influence on her of her visit to Norway was important. Actually meeting sober, good scientists who were convinced [of the seriousness of the issue].'[116] Most

---

[114] Dr G. Howells, interview.   [115] Sir Richard Southwood, interview.
[116] Rt. Hon. William Waldgrave, interview.

notable in this respect was the role of the work by professor Tamm on the so-called sulphur bank. Marshall argued he was persuaded after having 'spent the day' with him, 'looked through some reports' and then 'had dinner together' afterwards.[117] As he recalls the unpublished manuscripts 'filled in the missing gap'. Whereas Southwood argued that persuasion was essentially a process of gradual conviction, Marshall insisted all along on the instant persuasion after having seen Tamm's work: it was the missing link. It echoes the construct of the 'crucial experiment' that Merton already labelled as characteristic for Royal Society discourse of the seventeenth century.[118]

There is a perhaps even more striking example that illustrates both the ritualistic element of the employment of science and the embeddedness of scientific discourse in the British discursive style. Upon his return from Scandinavia Lord Marshall wrote 'a scientific paper': 'I sent this paper to Margaret Thatcher and said "In view of this scientific assessment, I now recommend that we spend the money."'[119] Here the manager of the electricity utility industry communicates with the Prime Minister, drawing on some sort of presumably meta-scientific language in which he could apparently express his argument for a major investment in pollution abatement equipment. Again science plays a ritualistic role, completely detached from normal scientific practice.

There are various other potential arguments that reinforce this reading of science as ritual, as an element of a specific discursive style. First of all, the massive investment required for the implementation of FGD would have run straight against the central commitments of the Thatcher government, which wanted to reduce government spending and avoid interference with industry. Furthermore, the fact that the Prime Minister attached much value to her own scientific background might have made it unlikely that she would have been convinced by those arguing that enough was known, especially while the Royal Society was still working on the issue. What is certain, however, is that after the Chequers meeting the results from SWAP had became a *conditio sine qua non*.

The influence of SWAP can also be recognized in the contributions of the eco-modernist discourse-coalition. Other quasi-

---

[117] Lord Marshall, interview.   [118] Merton 1970: 110 ff.
[119] Lord Marshall, interview; see also Marshall 1986.

governmental agencies (such as the NERC, ITE, NCC) found that their recommendations were all overruled by the fact that the Royal Society had not yet given its opinion. Furthermore, the involvement of the Royal Society also provided a counterweight to the HoC Environment Committee inquiry. In actual fact, the work of the HoC Environment Committee came a long way in creating a forum where a civil legitimacy of science could have been brought about. The traditional view on science was that scientists should acquire objective evidence and not express any opinion because this would invalidate their data. The HoC Environment Committee, however, explicitly asked for views and found that a majority thought that the state of knowledge legitimated action. In this respect the practice of parliamentary inquiries fulfilled a useful role in creating a civic forum for the discussion of acid rain as more than only a scientific issue. Nevertheless, its intervention came at the same moment at which the numinous legitimacy of science was reinforced through SWAP.

A special case is the positioning of the Royal Commission *vis-à-vis* SWAP. The Tenth Report of the Royal Commission on Environmental Pollution was entirely based on an eco-modernist footing, and did address the issue of the sort of science one needs in environmental policy-making. Yet it did not make strong recommendations on acid rain. The Royal Commission made the assessment that, in the face of SWAP, any call for action would be a lost cause. Yet the aversion against speaking out on the issue was at least partly due to the unfortunate double role of professor Sir Richard Southwood. He was an outspoken advocate of the application of ecological modernization to British pollution politics. Yet during his chairmanship of the Royal Commission he was unexpectedly asked, due to a curious course of affairs, to assume the chairmanship of the SWAP project as well. At the start of SWAP (1983) Southwood had been personally convinced that action should be taken. The result of his complicated personal involvement was that his focus was on creating an opportunity to make SWAP findings public in an intermediate report after two or three years. Hence as chairman of the Royal Commission he argued that the Commission did not want to speak out on the acid rain issue before at least some results from SWAP were known.[120] Yet the

---

[120] Sir Richard Southwood, interview.

unintended consequence was that the Royal Commission, one of the pillars in the eco-modernist coalition in the acid-rain controversy, could not be drawn upon to strengthen the credibility of the traditional-pragmatist stand. So it could come to pass that in its evidence to the HoC Environment Committee the CEGB positioned itself as committed to the 'spirit and intent of the Commission's recommendations' and the Government 'wholly accepted the Royal Commission's view' that high priority should be given to research.[121]

The analysis of the workings of the British science–politics interface in the acid-rain controversy demonstrates how the interpretation given to the SPA inhibited the application of ecological modernist ideas to the regulation of acid rain. Most notably the central role of the SWAP research prevented the application of the managerial stand that was advocated by the Royal Commission in its Tenth Report but had also been hinted at in Pollution Paper No. 11. Instead the acid-rain controversy shows how in actual practice the government avoided an open debate (both among scientists and between science and society) and reconfirmed the numinous legitimacy of science that had the stamp of the Royal Society at a time at which it had formally taken a more reflexive stand.

4.7. REGULATING THE REGULATOR

We are now entering an era where it would be almost politically unacceptable not to have FGD installed.

Chief Alkali Inspector, 1982

The Inspectorate have not required the installation [of FGD], because . . . they have regarded it as too expensive an imposition to constitute 'best practicable means'.

Department of the Environment, 1984

One of the pillars under the traditional-pragmatist approach in British air pollution regulation has always been the close consultation between the Inspectorate and the polluting industries. This materialized in the Victorian Era in the practice of Best Practicable

---

[121] CEGB in HoC Environment Committee (MoE) 1984: 6; DoE 1984*b*: 23.

Means (BPM). Since 1863 the Alkali Inspectorate has required industries to use BPM to avoid air pollution by statute. Since 1906 it gives the Inspectorate the duty to prevent the emission of noxious gases and, if this proves impossible, to render them harmless and inoffensive.[122] Yet while the adjective 'best' basically means 'most effective', 'practicable' is to be read as a pragmatic check that imposes limitations on the search for maximum environmental effectivity. Practicable was first defined by statute in the 1956 Clean Air Act: ' "practicable" means reasonably practicable having regard, amongst other things, to local conditions and circumstances, to the financial implications and to the current state of technical knowledge.'[123] This implies that some pollution might have to be tolerated even if a technical solution is available. For instance, if the costs would be unacceptably high. What is to be considered BPM is determined by the Inspectorate after consultation with the experts from the works involved. On that basis the Inspectorate regularly publishes so-called Notes on BPM that determine which pollution abatement strategy is considered BPM for the various listed noxious gases. SO_2 was first defined as noxious in 1974, but this excluded SO_2 emissions stemming from the combustion process itself and therefore power-stations were not covered by BPM. Until 1986 the abatement of SO_2 at power-stations was not seen as practicable because of the high costs of flue gas desulphurization.

BPM should be understood as a typical British alternative to the national emission standards that are found in many Western European countries. Yet there are also reasons to suggest that the actual effects of BPM are similar to other regulatory principles. BPM is broadly in line with the EC concept of Best Available Technology without Entailing Excessive Cost (BATNEEC) that formed the basis for the 1988 Large Combustion Plant Directive. Over the last twenty years the practice has undergone some marginal changes. The Royal Commission recommended in its Fifth Report the introduction of the principle of Best Practicable Environmental Option (BPEO) to replace the old principle of BPM in order to avoid cross-media pollution. The Government formally accepted BPEO in 1982. The term 'best practicable means' has

[122] *Alkali etc. Works Regulation Act* (1906), sect. 7, as discussed in Royal Commission on Environmental Pollution 1984: 45.
[123] Quoted in Royal Commission on Environmental Pollution 1984: 45.

still been regularly used, even in the years after 1982. For convenience sake we will therefore stick to using the term BPM to describe the established British practice of regulation.

Several institutional features indicate how BPM sustained the traditional-pragmatist approach. Firstly, BPM is a primarily administrative practice that keeps both politics and the public out. BPM is a story-line that defines a group of experts (both in industry and in the inspectorate) and works on the basis of a relationship of mutual trust and respect between these experts. It is a form of regulation that works best if based on consensus. One of the main features is in fact the avoidance of prosecutions. In this sense it is fundamentally at odds with the demand to lift the secrecy surrounding environmental decision-making voiced by actors such as the Royal Commission and the HoC Environment Committee. Secondly, BPM is to be understood as best practicable available means. The industry is under no obligation to develop new technologies. BPM thus fails to stimulate the invention and implementation of new pollution-abatement equipment. BPM is unlikely to be technology forcing.[124] Thirdly, 'economic incapacity' is a legitimate reason for inaction, which makes BPM both vulnerable to conjunctural economic change and to biased presentations of costs and the economic capacity of firms. Moreover, Vogel has drawn attention to the fact that even though the Inspectorate has to judge the economic capacity of firms to establish whether a technique can be regarded practicable, it employs neither economists nor accountants.[125]

It should not come as a surprise that BPM and the role of the Inspectorate has been heavily criticized in statements that drew on eco-modernist discourse. Here the confrontation between a national outlook as embedded in the 'proud record' of the Inspectorate, and the international outlook of the eco-modernists immediately becomes apparent. Sir Richard Southwood, chairman of the Royal Commission, saw foreign examples as the basis of his critique: 'We indeed had some contact with German environmental advisors at that time and we felt again that if this was the way that

---

[124] However, in a comparative case-study of the United States and Britain, David Vogel raises doubt as to whether this implies that BPM will result in slower progress towards pollution abatement than the more rigid and supposedly technology-forcing regulatory system of the USA: see Vogel 1986.

[125] Vogel 1986: 80, 83.

perhaps the most successful country of Europe was going, perhaps we should take a somewhat parallel route.'[126] The modernization perspective implied that actors were looking to the most advanced industrial societies. If environmental regulation was to be a positive-sum game then only a leading role by the government could help bring this about. Waldegrave recalled:

I developed a theory on the basis of my observation that the more highly developed a society was, the more concern . . . was shown for environmental matters and that therefore the most highly developed markets in the world were also the most likely to have the highest environmental standards. And that if a society was trying to modernise industry, which was high on our political agenda, to make a function change in the capacity of out industry, one should be linking oneselves to these highly developed markets and that Britain was still somewhat inclined to look at traditional markets which worked the other way round.

Similarly in an eco-modernist argument British environmental regulation should become technology forcing which brought conflicts with the established BPM practice.

When I first arrived I had to make a decision on the removal of lead from decorating paint. The department said that we should give industry five years to make the change, I said why not 18 months? The industry immediately agreed which made it perfectly clear to me that they could have done it in six months. So one learned all the time, that with a high tech industry, basically ICI, things were often moving faster than the bureaucracies understood . . . I became very convinced, particularly after the Munich conference [July 1984], with the idea of standard setting as an industrial moderniser.[127]

This view was in fact confirmed by the Inspectorate itself. In 1982 it argued that the breakthrough in the technical difficulties with the FGD process was 'enforced by legislation in Japan and America.'[128] The story-lines of innovative foreign alternatives led to the wish on the part of eco-modernists to change the workings of the Inspectorate. As early as 1975 the Royal Commission had

---

[126] Sir Richard Southwood, interview.
[127] Rt. Hon. William Waldegrave, interview.
[128] HoL Select Committee for the European Communities (MoE) 1982: 42. As will become obvious below, the Inspectorate were remarkably open during these 1982 hearings. It might well be that this was caused by the fact that the acid-rain issue was not a very central political issue at the time. Political attention would most likely have caused a more cautious performance by the Inspectorate.

argued in favour of a unified pollution inspectorate. But within the eco-modernist frame this recommendation was strengthened by the wish that this inspectorate should become more independent from government. Waldegrave's position was that 'I wanted to establish an independent inspectorate. I wanted a voice that was clearly independent of government to change the beliefs and influence the atmosphere. I thought there was an educative role for the inspectorate as well as a purely functional role.'[129]

Hence despite its conservative role within the traditional-pragmatist way of running environmental protection, the Inspectorate was clearly recognized as a potential broker for ecological modernization. This was also an essential component in the *Environment Policy Review*, an internal DoE document written in 1985 by the Central Policy Planning Unit and commissioned by Waldegrave.[130] However, although there seems ample evidence for the fact that the top of the DoE was inspired by the ideas of ecological modernization, Waldegrave was only partially successful in reorganizing the reactive practices that dominated the everyday workings of the department. Of course, the government erected a unified pollution inspectorate in April 1987.[131] Yet it was heavily underfunded and understaffed and lacked any substantial new powers. It was described by eco-modernists as 'only half a loaf' (Southwood).

The institutional analysis that we have presented so far should not lead to an interpretation of the BPM practice and the Inspectorate as structurally tied to a pragmatist approach to environmental regulation. In actual fact the effect of the regulatory practices was much more open and was dependent on the meaning given through various discursive interchanges. This is illustrated in the acid-rain case. How BPM should be put to work was to be determined through an interactive process of positioning. First I will consider the Inspectorate's interpretation of its own role and

[129] Rt. Hon. William Waldegrave, interview.

[130] Central Policy Planning Unit 1985. It also proposed to extend the terms of reference and title of the Royal Commission on Environmental Pollution and suggested considering re-establishing the Commission on Energy and Environment. Furthermore it presented a 'skeleton White Paper' that could serve as a basis for a White Paper in 'mid-1986.'

[131] For the history of the unified inspectorate of pollution, see O'Riordan and Weale 1989; for a first assessment see Gibson 1991.

position, followed by a discussion of the way in which this related to various external factors.

One of the principal qualities of BPM is that it leaves much discretion for the Inspector. This also implies that they have to constantly rethink what BPM should be. The acid-rain case shows that this can result in differing interpretations even among the Inspectors. Mr M. F. Tunicliffe, Air Pollution Inspector, defined 'harmless' as the key word and argued: 'We have never gone as far as saying harmless to the lichens and little butterflies, I mean how far do you take harmless down? But certainly we have always said if public health was implicated then costs were no matter.'[132] It is easy to see that this interpretation of BPM works against an intervention in the acid-rain debate. However, while giving evidence to the House of Lords in 1982, Dr L. E. Reed, then Chief Air Pollution Inspector, argued that 'practicable' was the key word. Damage would be a good basis for action, he argued, but the question really was whether solutions would be practicable (i.e. whether they did not involve excessive costs).[133] Interestingly, Dr Reed maintained as early as in 1982, that FGD was technically available. He contended that if new power-stations were to be built, FGD should probably be considered, even if there was no acid rain at all.[134] But how did he come to this conclusion? He argued that there was 'no apparent reason why we should, *for the sake of the United Kingdom air quality*, require installation on existing power stations.'[135] Yet Dr Reed also argued that 'We are now entering an era where it would be almost *politically* unacceptable not to have FGD installed.'[136] In the same hearings the Inspectorate also used as a consideration the fact that the UK was about to be one of the few countries that did not use FGD. This indicates that the definition of 'practicable' could also be influenced by political considerations. Hence BPM calls for political judgement too. Or, as Tunicliffe put it: 'In making judgements you are aware of what society's views generally are.'[137] What is more,

---

[132] Mr M. F. Tunicliffe, interview.

[133] HoL Select Committee on the European Communities (MoE) 1982: 40.

[134] Ibid. 40. Yet the Inspectorate considered the retrofitting of FGD too expensive and therefore FGD did not become BPM.

[135] Ibid. 41, emphasis added.   [136] Ibid., emphasis added.

[137] Mr M. F. Tunicliffe, interview.

whereas Vogel illuminated that the Inspectorate was supposed to judge economic incapacity while it did not employ economists or accountants, this suggests that the Inspectorate draws on its numinous legitimacy as an expert body to make political judgements too.

The fact that the everyday discursive creation of BPM makes it far from being an insulated administrative practice that relies on expert judgement, becomes even more apparent if we examine the role of the Inspectorate in the September 1986 decision to retrofit three power-stations with FGD. With the 1986 decision the Inspectorate did not set the standards but followed the innovations suggested by the electricity industry it was supposed to control. As we saw above, the 1986 decision to retrofit power-stations was recommended to the government by the CEGB and was then subsequently approved by the Treasury and announced by the DoE in London and Mrs Thatcher in Oslo. In actual fact FGD was not considered BPM until after both the government and the industry had agreed on the installation of FGD.

Hence, even though the Inspectorate had the legal authority to describe FGD as BPM, it never did. In the traditional reading one would argue that the Inspectorate made its own assessment that the benefits to nature did not outweigh the costs to industry. However, it seems more likely that the decision revolved around other variables. Tunicliffe recalled that the Inspectorate made the decision dependent on the developments in the EC negotiations over that Large Combustion Plant Directive:

One had to tread very carefully because since the government had a reserve, a general reserve, its own Inspectorate could not be busy or publicly busy, undermining that position . . . The Inspectorate's view had to be aligned with Government because of the great implications . . . the Inspectorate consciously let the national negotiations take precedence . . . The Inspectorate therefore kept its head down, saying 'we aren't sure yet'.[138]

So the conclusion has to be that the Inspectorate had the legal powers to facilitate problem closure but did not employ these because of its self-positioning *vis-à-vis* other institutes. More specifically, it reflected the impact of the representation of the social reality of the traditional-pragmatist discourse-coalition.

[138] Mr M. F. Tunicliffe, interview.

It can be shown how the meaning of the practice of BPM in this case was actually the product of an intersubjective discursive positioning. Speaking on the acid-rain issue in 1984 the government argued that:

flue gas desulphurisation is a proven technology for removing sulphur dioxide from power station plumes but the . . . Inspectorate have not required its installation, because at a capital cost of some 150 million pounds for each major power station, they have regarded it as too expensive an imposition to constitute 'best practicable means'.[139]

Here the Inspectorate is positioned as the legitimate and independent authority whose expert judgement the Government would follow. The same positioning-game was played by the CEGB. It positioned the Inspectorate as the legitimate and powerful experts and subsequently hid behind their judgement, arguing that the Inspectorate would take a precautionary approach if FGD would not require an 'undue financial burden'.[140] The suggestion that was made is thus that the Inspectorate will set the standards. If it does not require FGD, this is used by the positioners as an expert argument against popular pressure.

In actual fact the Inspectorate itself consciously tried to avoid showing any independence. In the context of the acid-rain issue that was considered to be totally inappropriate:

MAH: But you know that Angus Smith at some point went beyond his brief and recommended . . .

Tunicliffe: Oh yes, but acid rain became political straightaway. And from that point of view, the Chief Inspector's view, the Inspectorate's view had to be aligned with government thinking and policy because it clearly had big national and international implications.

However, instead of putting all the blame on the plate of the Inspectorate and accusing it of craven behaviour, it seems to make more sense to understand the interpretation of the BPM practice in the context of the general discursive play on that practice. BPM was evidently very much part of the core of the traditional-pragmatist approach to environmental pollution. Its historical roots in the Victorian Era made it one of the prime objects of the proud

---

[139] DoE 1984*b*: 2.
[140] HoC Environment Committee (MoE) 1984: 4; see also HoC Environment Committee (MoE) 1988: 39.

record narrative. Furthermore it was this metaphor that eco-modernists did not really dare to contest. The HoC Environment Committee consciously aimed to show that it worked within the frame of BPM, using the appropriate language in its Acid Rain report. The Royal Commission also emphasized the advantages of the British practice of BPM and its suggestions for innovation were of the most pragmatic sort.[141] Its concerns could be captured in the eloquent words of Lord Sherfield: to ensure that 'the best practicable means do not degenerate into the cheapest tolerable means'.[142] In doing so, it still helped to keep up BPM as traditional-pragmatist practice.

In all, BPM supported, in a paradoxical way, the political primacy in pollution politics while the practice derived its fame and credibility precisely from the fact that it left pollution abatement to the discretion of experts. Indeed, this was the way in which politicians legitimized their continued trust in the BPM practice that some critics thought ineffective and an impediment to an innovative environmental policy.

4.8. CONCLUSION: SCIENTISM AS DISCURSIVE BIAS

The British case-study shows how ecological modernization started to conquer the British environmental domain from the early 1980s onwards. In light of the ideas of the influential *World Conservation Strategy* the well-established *post hoc* policy approach emerged as an anachronism working with dated problem perceptions and equally dated ideas about environmental management. From an eco-modernist point of view the traditional-pragmatist discourse, that was firmly rooted in the early air-pollution control strategies of the Victorian Era, seemed to leave the British Government ill-equipped to deal with the new generation of environmental issues that had been outlined in the WCS. We have seen how these new cognitions appealed to some industrialists, who became attracted to the idea of an integration of environmental concerns in industrial restructuring. The top of the Department of the Environment became interested, as well as the Royal Commission on

---

[141] See for instance Royal Commission on Environmental Pollution 1976, 1984.
[142] Lord Sherfield quoted in Royal Commission 1984: 41.

Environmental Pollution, the HoC Environment Committee, NGOs, and various quango organizations like the NCC and NERC. Especially the evidence given to the HoC Environment Committee shows convincingly that acid rain fulfilled the role of catalyst in this debate on policy discourses. Yet the study of the acid-rain controversy at the same time strongly suggests that although the eco-modernist ideas might have been well received in the abstract, the emerging discourse-coalition of ecological modernization nevertheless failed to impose its logic on the actual decision-making on acid rain. Instead, the traditional-pragmatist discourse-coalition, with the CEGB, the Royal Society, and the Cabinet as identifiable key protagonists, that could rely on many well-institutionalized practices, stretched the traditional policy discourse to cover the new issue, thus denying its uniqueness and claiming a continued legitimacy of the old regulatory arrangements.

In all, my argumentative analysis of the acid-rain controversy does not lead to a radical refutation of the general points about the British style of regulation made by authors like Jordan, Richardson, Vogel, Macrory, or Weale.[143] For instance, there was ample evidence that bargaining was more important than imposition and consultation with client groups outside Government most certainly meant that initiatives basically had to be cleared. Similarly, there was ample evidence of the importance of the existence of certain expert networks that were often more a coalition of certain departments with their respective client groups than interdepartmental constructs. What I have shown is, first of all, the more fundamental meaning of the acid-rain controversy. The discursive construction of this issue decided whether or not the routinized style of regulation could still be seen as legitimate and sufficiently effective in face of the sort of challenges of the environmental problems posed in the 1980s. Argumentative discourse analysis thus opened up a new emphasis in the analysis of environmental policy-making. Secondly, and related to the above, it has shown specific practices through which these features are discursively (re-)produced in the environmental domain in the first place. In this concluding section I will first show the concrete cases where ecological modernization led to institutional change.

[143] Jordan and Richardson 1987; Richardson and Jordan 1979; Macrory 1986; Vogel 1986; Weale 1992.

Subsequently we will discuss the reasons why in the end ecological modernization failed to put its mark on the regulation of the acid-rain controversy.

Ecological modernization might not have dominated the thinking about concrete solutions for the acid-rain problem, but it surely did change the debates on the task for environmental policy-making and, indeed, also influenced some institutional routines in the context of the discursive construction of acid rain. It was precisely in various remote pockets of the environmental domain that the discourse of ecological modernization actually changed existing institutional practices. A first example is the institutionalization of a more comprehensive and permanent air-quality monitoring network. This implies an implicit acknowledgement of the new structural nature of environmental perturbations. Most certainly, the stimulus for this change did not come through the acid-rain debate alone. International political developments, especially at EC level, reinforced the arguments of the domestic call for better monitoring. A second example of the institutionalization of the discourse of ecological modernization was the acceptance of the multiple stress hypothesis by the Forestry Commission and, the government's acceptance of the notion of critical loads. Both practices reflect the influence of the precautionary principle. This eco-modernist policy axiom can be seen as a fundamental break with the prevailing SPA practice. The case-study showed the distinct reasons that NGOs, biologists, pollution scientists, politicians, and press may have to support the precautionary principle. The precautronary principle starts from an assumption of the serious damage done to the environment and is sceptical about the likelihood that nature could ever be fully understood by taking its elements apart. Although this might not have been immediately apparent in the acid-rain controversy, this endorsement of the precautionary principle will most certainly have major consequences for the operational routines of the Science-based Policy Approach, sketched in this chapter. Most likely it will not lead to a total abandonment of the SPA, but to an adjustment in the direction of a more regulatory sort of science in the sense described by Jasanoff. It would thus enhance the legitimacy of the claims made by scientists who are active in pollution science, claims that were refuted as unscientific in the acid-rain controversy.

An important practice in the proliferation and structuration of ecological modernization was undoubtedly the 1984 HoC Environment Committee Inquiry. It was an example of a practice that enhanced the reflexivity in the environmental domain. It brought together the wide range of actors working on the issue that could have allowed for a more open alternative to the expert led Science-based Policy Approach. It produced the most elaborate exchange of ideas, of scientists, industrialists, and experts alike. Likewise the various Review Groups started by Waldegrave were a positive attempt to create institutional settings for reflexive interdisciplinary debate. Nevertheless, overall ecological modernization failed to put its mark on the regulation of the acid-rain problem. As we have seen there were some clearly identifiable practices through which this power effect was created. Here we will not reiterate these findings but will try to infer some general features of the environmental domain from the analysis of the practices through which acid rain was regulated instead. After all, acid rain was not just another issue, it was perhaps the most important environmental issue of the early 1980s and most certainly the issue against which ecological modernization was put to the test.

The first characteristic feature of the British discursive style is its antagonistic positioning. The acid-rain controversy presented two totally distinct story-lines that reflected two rival policy discourses. They related to one another in an almost exact thesis–antithesis format. Table 4.1 illuminates that the dissimilarities of the two story-lines covered immediate aspects of the acid-rain problem, such as the damage caused, the attribution of responsibility, and the appropriate way to come to regulation of the problem at hand, but included broader questions as well. More specifically, the new eco-modernist story-line on acid rain challenged the legitimacy of emissions, the authority of specific experts, and the credibility of specific decision-making structures. Yet despite the fact that the reconstruction of the acid-rain debate indicates the wide scope of the controversy, the actual discursive challenge posed by the eco-modernist discourse-coalition was much more limited than the ideal typical reconstruction of the story-lines suggests.

The first comment on this table should certainly stress the importance of the structuring power of the external force of the dual party system which itself is reinforced by various institutional practices. This lack of a tradition of coalition cabinets reinforces the antagonistic discursive style. Yet within the confines of the

TABLE 4.1. *The discursive space of the British acid-rain policy debate*

Traditional pragmatism	Ecological modernization
Positioning of acid rain as a policy issue	
– acid rain as an SO_2-related problem, damage is foreign affair	– acid rain as an SO_2-related problem but damage is here too
– national specific: 'proud record'	– understanding of record in international terms: UK as 'dirty man of Europe'
– quasi-scientific inductive understanding of acid rain derived from *ad hoc* analysis of isolated pollution incidents. Implicit belief: 'no crisis'	– deductive understanding of acid rain in terms of the ecological crisis; acid rain as illustration of the 'cumulation of gases in the atmosphere' (WCS); explicit belief: structural pollution
– acid rain as serious incidental issue	– acid rain as programmatic issue
– 'proud record', i.e. acid rain as residual problem	– 'proud record' is anachronism: regulation of new problems requires institutional change,
– acid rain to be understood in homogeneous air-pollution discourse	– i.e. acid rain calls for heterogenization of air-pollution discourse
– acid rain as case of irresponsible scaremongering	– acid rain as potential focus for political disillusionment
Repercussions of acid rain for the policy domain	
– because of 'proud record': no need for new policy discourse	– because of seriousness of issue, need for new policy discourse
– need to prove damage to impose constraints	– because of seriousness: reversal of the burden of proof: 'guilty unless proven innocent'
– pollution is as such legitimate	– pollution is, in principle, undesirable
– against scaremongering	– against 'complacency'
– strong individualist commitment	– reluctant integration of structural dimension
– emphasis on tradition: no explicit reference to neo-liberal commitments	– explicit attempt to position environmental care within neo-liberal politics (Waldegrave, DoE Policy Review)

Science and expertise

- 'policy in danger of running ahead of science'
- imposition of Royal Society as norm for pollution science: government internalizes validity criteria of research science (understanding causal chain required)
- emphasis on need for precise study of cause–effect relationships
- primarily scientific legitimacy
- Inspectorate positioned as independent authority
- Royal Society as ultimate authority above the sphere of pressure-group politics

- 'cannot wait until all evidence is in' (Rossi)
- understanding of pollution issues necessarily limited by complexity of eco-systems and impossibility of controls (correlation suffices)
- need for new vocabulary: multiple stress, critical load, precaution, emphasis on research and development of solutions
- need for civil legitimacy: both measured and perceived risks are important
- emphasis on need for reorganization to create independent and unified Inspectorate
- Royal Commission as respectable mediator of public concerns

Confines of environmental regulation

- environment as sector concern
- regulation as a techno-administrative affair
- act only if experts show it would solve real environmental problems in an effective and economically feasible way
- avoid unnecessary expenditure
- control negative externalities
- justice—utilitarian: is it just to use taxpayers' money to resolve foreign problems?
- problem closure: research needed
- social accommodation: reinforcement of past achievements, education, emphasis on scientific authority
- do just enough, just in time

- environment as integrated concern
- regulation as a socio-political affair: taking public feeling into account
- act now: environment requires action
- avoid irreparable damage to nature
- internalize costs
- distributive justice: is it just to harm innocent victims?
- problem closure: FGD is the obvious solution
- social accommodation: abate largest emissions, remove obsolete bar on disclosure of information, take public seriously
- install FGD to be able to join 30% Club

environmental domain, discourse analysis illuminates processes that give an insight into the way in which such a structure is reproduced. The case-study showed, for instance, the extent to which the dynamics of story-lines were related to time-and-space-specific practices rather than actual individuals. Some of the key protagonists of the eco-modernist coalition, most notably minister Waldegrave and his chief scientist Holdgate, but to a lesser extent also the HoC Environment Committee and the Royal Commission, were seen uttering different statements depending on their time-and-space-specific positioning. Waldegrave and Holdgate most certainly personally supported the cause of ecological modernization, yet failed to secure enough support on the Cabinet level to alter the official Government story-line. They did make some headway within the DoE and had considerable success in organizing support for the abstract eco-modernist policy principles both within industry and in circles of environmental NGOs.[144] However, Waldegrave and Holdgate were held to defend the governmental line at all official meetings, whether these were HoC Environment Committee hearings or international ministerial conferences on acid rain. These formal restrictions most certainly did not apply to other protagonists of the eco-modernist discourse-coalition, such as the HoC Environment Committee or the Royal Commission or indeed Friends of the Earth. Nevertheless, they did not unequivocally frame the acid-rain problem in eco-modernist terms either. This practice of arguing one's case at key moments in the terms of the discourse of power is the second feature of the British discursive style. The classic analysis of the British two-party system suggests that the only way to make headway is to try to seek incremental adjustments within the frame used by the dominant political bloc. But one may wonder on what idea of politics this conviction is actually based. If politics is the argumentative struggle in which actors seek to impose their definition of reality on others, the question is to what extent government can simply impose definitions or whether it has to organize its authority in a social

[144] For the business involvement see the CBI statements in the HoC Environment Committee MoE of 1984, and the debates held at the meetings and conferences of UK CEED (e.g. the Conference on *Environmentalism Today—The Challenge for Business* in April 1986 (UK CEED/NERA, 1986)). The support of the NGOs was secured through the initiation of regular round-table meetings with some members of the professional NGO élite such as Tom Burke (Green Alliance), Robin Grove-White (Council for the Protection of Rural England), and Tim Wilkinson (FoE).

debate. If the latter is the case, the easy reference to the British two-party system does not suffice. Then one has to investigate how a government manages to assemble credibility and temper the power of alternative claims.

Here there are some interesting mechanisms of discursive power formation at work. It seems that the discourse of ecological modernization was not excluded from the environmental domain as one would exclude an opposing political ideology. There was no gatekeeper who intended to keep out certain ideas. Indeed, the Government suggested it was rather sympathetic to its main ideas. Yet there are several subtle mechanisms that nevertheless effectively limited the influence of ecological modernization on the regulation of acid rain. Somehow, actors in some settings simply refrained from framing their concerns drawing on eco-modernist concepts. Was this a matter of what Friedrich called the 'anticipated reaction' to the 'Machiavellian guile or force' that Voltaire already thought was the characteristic flip-side of the highly praised British tolerance?[145] It seems it might have been slightly more complicated. First of all, the reproduction of the traditional-pragmatist approach was routinized in the workings of all kind of institutional practices. Actors who thought that eco-modernist terms made most sense of what was going on in the environmental domain faced a discursive dilemma. In effect they constantly had to make a choice whether to also challenge the prevailing traditional-pragmatist structures or to operate (think, speak) within the existing formats. The discursive dilemma existed in the fact that either choice came at a considerable cost. For instance, they could have insisted that acid rain should be understood as a metaphor. Acid rain would appear as the more or less coincidental first issue to emerge out of the general problem of structural pollution. This would legitimate, indeed call for, a new approach to environmental policy-making. In that case they would have had the freedom of framing the issue (the problem definition, the relations with solutions, the causes, the costs and benefits) and might have been more successful in illuminating the shortcomings of the prevailing institutional routines in public discourse. Yet most likely they would have had considerable difficulty in getting through to decision-makers. Alternatively, if the eco-modernists were to choose to go

---

[145] Crick 1991: 93.

along with the prevailing framing of the issue or only to problematize certain aspects they would most certainly have more chance of being heard and understood. Yet this would come at the cost of having to persuade others in a discursive structure that was not furnished to define the problem in the contextual sense that was characteristic of ecological modernization.

Whether it was a case of an anticipated negative reaction to an eco-modernist framing of the acid-rain issue, or the result of careful strategic deliberations, the empirical research demonstrates that at key moments protagonists of the eco-modernist discourse tried to defeat the traditional pragmatists at their own game, rather than seek to change the terms of the debate. The extent to which actors complied with this respect for the rules of the game is easily overlooked, since it was at odds with the publicly uttered eco-modernist story-line. Whether it was in setting out the acid-rain problem at Chequers, in making recommendations in the Tenth Report of the Royal Commission or in the HoC Environment Committee Reports, eco-modernist concepts were sometimes almost absent in the acid-rain policy debate even though the discourse of ecological modernization was rapidly replacing the traditional-pragmatist notions both in the international discussions and in the domestic theoretical debate on policy styles. The EC had already accepted the precautionary principle in its Environment Action Programmes, the OECD had made ecological modernization into the core of the Ministerial Conference of July 1984, and in Britain the government had thought it appropriate to amend its definition of the Science-based Policy Approach. Yet at key moments eco-modernist reasoning was simply absent in the actual acid-rain debate.

The explanation of this phenomenon is rather complex. Throughout this chapter I have shown how one actor may utter different points of view depending on the practices in which he or she is engaged. In this sense the clear-cut distinction of two antagonistic story-lines also leads the discussion somewhat astray. The ambiguous and contradictory nature of individual utterances provides us with at least an essential correction of previous insights into the British discursive style. The antagonistic story-lines are thus not to be seen as tied to persons but are characteristic of the practices of the environmental domain.

In other, less publicly visible, discursive practices the acid-rain

controversy had markedly different features. Nearly all participants in the public acid-rain controversy had a rather precise implicit
understanding of the structured ways of arguing in the corridors
of power. What is more, the case-study clearly shows that actors
tend to take these implicit understandings as guidelines for their
interventions. We can see how committed actors who helped to
propagate eco-modernist ideas in other contexts time and again
conformed to the old standards of credibility, failed to alter the
terms of the debate, and failed to employ the eco-modernist concepts which they elsewhere maintained had a greater cognitive
potential and heuristic value. Here they seem to have been guided
by more than opportunistic strategic considerations alone. There
seems to be an implicit assessment of how interventions ought to
be made, and how they are to be made effectively. Hence it seems
to point to the existence of a certain shared understanding of a
discursive style in policy-making from which eco-modernists failed
to escape. The radical NGO Friends of the Earth emphasized that
they thought it better to argue a case by briefing the conservative
Bow Group instead of speaking out themselves.[146] The Royal
Commission did not want to challenge the authority of the Royal
Society,[147] the HoC Environment Committee were afraid of 'getting their science wrong'.[148] In many instances eco-modernists
seemed to have been very concerned to show they were 'the right
kind of people'. Yet the unintended consequence of playing by the
rules is to have a discursive logic imposed on you.

Here the respect for the rules of the game was closely related to
the third and perhaps most influential feature of the British discursive style: the unchallenged centrality of scientific discourse in policy
deliberations. Moral arguments (that were, for instance, frequently
raised and discussed in the HoC Environment Committee hearings) did not find their way into the actual policy-making-oriented
interventions.[149] Even the economic dimension, for example,
through an emphasis on the positive-sum format (the suggestion
that acid rain should be seen as a stimulus to innovate production
processes), was omitted from the actual policy debates, although
it was frequently found and generally endorsed in the abstract. In

---

[146] Mr C. Rose, interview.
[147] Sir Richard Southwood, interview.   [148] Sir Hugh Rossi, interview.
[149] Cf. e.g. the hearings with the Fourth Report (HoC Environment Committee, 1984).

the section on SWAP we have given ample attention to the dominant role for science. Indeed, one could argue that the first two
features of the British discursive style as described above (the
antagonistic positioning and the respect for the rules of the game
or the discourse of power) were largely argued out by drawing on
scientific discourse. This tendency for scientism, in which political
discourse is contaminated by pseudo-scientific concerns (be it in a
much more complicated way than by using science as fig-leaf) may
be seen as the dominating feature of the British discursive style in
the acid-rain controversy.

We have already shown that it was in fact a particular exclusive
sort of science that formed the backbone of the discourse-coalition
of traditional pragmatism. Most certainly the 'reluctance to read
too much into the evidence' of traditional pragmatism was itself
not an incremental feature as the fig-leaf metaphor suggests. As
Merton and others have shown, the empiricism of the experimental approach was well rooted in the Puritanism of seventeenth-
century England.[150] This well institutionalized and highly respected
research science started from the premiss that there was an order
in nature, and an order that was perfectly intelligible. This assumption was perhaps most eloquently criticized by Whitehead
who saw this as 'an unconscious derivative of medieval theo-
logy.'[151] The point here is that the two discourse-coalitions differed
over fundamentals like what constituted knowledge. And it was
not surprising that this question could not be resolved in the context
of the acid-rain controversy.

It is undeniable that this structural tendency towards scientism
was reinforced by the external influence of the political ideological
commitments of Thatcherism. The installation of FGD would increase public spending at a time when this was under strict control.[152] Likewise it could have been read as an interventionist act
(unnecessarily?) restricting the freedom of operation of industry.
One should not forget, however, that the environment had not
been a prominent issue in the previous decades either. Keynesian
welfare-state concerns had dominated the post-war era and the
environment had only become topical with the smog issue in the
1950s and the general concern in the early 1970s. In this sense it

---

[150] Merton 1970; Shapin and Schaffer 1985.
[151] Quoted in Merton 1970: 108.   [152] See Weale 1992: 87.

seems more important to emphasize the extent to which the Thatcher government drew upon a structural feature of the British policy style. Neo-liberal ideas gave this non-interventionism in the environmental domain new meaning. Precisely here we see to what extent Waldegrave and the Royal Commission went against the political forces that upheld the discourse-coalition of traditional pragmatism. Waldegrave might have been a minister (and later became a member of Cabinet) but his ideas to combine Thatcherist commitments about industrial restructuring with ecological modernization clearly put him outside the influence of what Weale calls the 'libertarian conservatism' that dominated the Cabinet.[153] Similarly, the Royal Commission argued for an active governmental stand to make the positive-sum game work in its eco-modernist Tenth Report.

In this context it is important to emphasize the broader implications of the seemingly incidental decisions in the context of the acid-rain issue. If we accept that the acid-rain controversy was not about technical facts alone, but was a potentially programmatic issue, in the context of which a whole policy discourse was being judged, then the meaning of these seemingly incidental decisions has to be seen in a different light. They were in fact the transfer points where the credibility of old and new ideas were to be assessed. Hence in these decisions much more was at stake than the choice of technical solution A or B or time-frame Q or R. Friends of the Earth, for instance, did not really involve themselves in debates on policy styles. It was 'beyond them'. Yet they thus failed to argue what they privately believed, namely that the policy discourse should be altered precisely because of issues like acid rain. Indeed, they thus helped to reproduce the empiricist orientation that upheld the traditional-pragmatist policy discourse. The choice of the Royal Commission not to challenge SWAP was in fact a reinforcement of the suggested hierarchical relationship between research science and pollution science and thus denied the latter the status of an independent and increasingly relevant domain. The collective decision not to challenge the central role for scientists reinforced the SPA and inhibited the development of new institutional structures through which a civil legitimation of science could be achieved.

[153] Ibid.

The symbolic dimension of the policy process is not unimportant in this respect. Certainly one of the most remarkable facts about the British acid-rain controversy is the combination of its organization around two antagonistic story-lines whilst at the same time the very definition of the issue was not contested. Admittedly, one story-line held that there was no problem while the other argued that there was a very serious problem. Yet all actors settled for a definition of the acid-rain problem as an SO_2-related problem (thus excluding NO_x, NH_3, and ozone), positioning the CEGB as main culprit (excluding other industries, traffic, and agriculture), and presenting FGD as the obvious solution (excluding using low sulphur coal, ignoring the possibility of reducing car traffic, and, especially, giving less weight to reducing demand through energy conservation).[154] What is more we have seen that the CEGB was rather quick to shift its position and to admit that 'it was probably guilty of killing fish'. The obvious effect was that it made the controversy between the eco-modernist and traditional-pragmatist discourse-coalitions look like an incremental quarrel. How can you play out a fundamentally different policy discourse if all it boils down to is having FGD installed more quickly? This arbitrary confinement of the issue thus made the problem appear as a far less convincing example of a programmatic issue requiring a radical change of policy regime. If it could be solved by installing some scrubbers, why all the fuss?

All through the acid-rain controversy FGD featured as the proven solution. There can be little doubt that this hindered the potential to draw on the acid-rain example to push a new policy discourse. FGD was a solution with great symbolic appeal. It had the right degree of immediacy, directness, and concreteness to appear as an appropriate response.[155] Placing scrubbers on a chimney is a clearly

[154] Admittedly, energy conservation was an integral part of the FoE critique and was underscored by the HoC Environment Committee in both its 1984 and 1988 reports. Yet this would have been a much more appropriate solution if the problem had been presented in a more structural frame. The present framing precluded a meaningful debate on solutions whereas a structural problem would have called for a structural solution, i.e. energy conservation.

[155] This was to a large extent caused by the historical frame of reference in which both discourses sought to understand acid rain. FGD was first and foremost a response to the tall stacks of the past. Now it was known that the strategy of 'dilute and disperse' did not work. The new wisdom was 'what goes up, must come down'. The popular image of FGD was that of a filter put on a chimney, which certainly seemed an appropriate response.

defined act which can be interpreted as a sign of success by followers of both story-lines. Critics could point to change[156] and the traditional pragmatists could claim that the legitimacy of their discourse still held.[157] Electricity prices would go up slightly but not enough to worry anybody. In this sense FGD was the classic example of a technological fix that, although it might be expensive and inefficient,[158] is still the preferred regulatory option since its application does not upset the social equilibrium.[159] Yet it is certain that by favouring FGD the eco-modernist helped reproduce the dominant single problem–single answer construction of the environmental problem. They failed to put the more radical eco-modernist alternatives on the table and thus failed to use the emblematic issue of acid rain as a lever to open up existing decision-making routines. What is more, they failed to position acid rain in the context of the decision-making alternatives they favoured elsewhere (although sometimes within the very same cover, like in the case of the Tenth Report of the Royal Commission).

In the context of this symbolic dimension one should also consider the semiotics of traditionalism within the traditional-pragmatist discourse-coalition. It constantly referred back to its roots in the Victorian Era. Its Alkali Inspectorate, Best Practicable Means, the reference to Angus Smith who had discovered acid rain way back in 1852, the reiterated proud record, all helped to suggest that the traditional-pragmatist practice was a permanent, natural regime. This reified style of regulation made it not a historically specific, transitory state of affairs, that, if circumstances changed, might have to be altered or might even have to make way for a more effective approach. SWAP strengthened the ethos of the traditional approach by bringing in that other icon of virtuous British traditions, the Royal Society, 'probably the most prestigious scientific academy in the world'.

[156] As indeed they did, see HoC Environment Committee 1988.
[157] As indeed they did, see DoE evidence to HoC Environment Committee 1988.
[158] The preferred FGD process uses limestone as feedstock and causes huge cross-boundary pollution in the form of low-quality gypsum.
[159] In this context one may also underline the fact that the CEGB consciously refrained from trying to reopen this arbitrary confinement. The CEGB realized that a call to bring in NO_x and ozone, let alone an intervention in which the extension of the nuclear programme would be put forward as solution for the problem of acid rain, would only be read as trying to divert attention away from power-stations and would thus endanger its already patchy reputation as 'trusted actor' (Lord Marshall, interview).

Several of the reported new eco-modernist practices, then, might in the end come to have considerable discursive power. However, for the British acid-rain controversy the effects were rather limited. The key reason here was that the knowledge input for the regulation of the acid-rain problem was essentially controlled by the SWAP project. Hence the discourse-coalition of ecological modernization most certainly left its mark on the regulation of the controversy. Just as it seems certain that the 1986 decision would not have been taken if the SWAP project had not reported, it seems fair to say that SWAP would not have taken place if ecological modernization had not emerged as a more radical challenge to existing policy commitments. At the same time there seems little doubt that the inauguration of SWAP was undoubtedly the single most important discursive interpellation in the acid-rain controversy. Likewise, it is clear that the dominant role for SWAP precluded the application of eco-modernist ideas so as to make environmental politics more reflexive and more progressive. It deferred decision-making, closed the discursive standards, depoliticized the issue, and precluded any open debate both among scientists or in even larger fora. Key to this seems to have been the involvement of the Royal Society. It was the ethos of that institute (perhaps even more than its actual agency) that helped the Government to withstand pressure stemming from the remarkable fact that a large majority of scientists and experts actually challenged its acid-rain stand. In this sense SWAP not only obstructed an increase in reflexivity of the policy process. In actual fact it meant a turning back of the clock.

5

The Micro-Powers of Apocalypse: Ecological Modernization in the Netherlands

5.1. INTRODUCTION

Environmental policy-making in the Netherlands constitutes one of the most interesting examples of ecological modernization, illuminating some of its most fundamental opportunities and problems. At the same time the history of the Dutch policy planning is little known. The Dutch 'policy planning' approach is one of the most comprehensive attempts to integrate the ecological dimension in governmental policy but has, as we will see, its own virtues and problems. This chapter presents a detailed account of this approach, showing the interaction between the emergence of eco-modernist discourse on the one hand and the regulation of acid rain on the other. This will show how the understanding of the politics of environmental discourse requires more than the examination of policy documents and story-lines alone. In the Dutch case eco-modernist story-lines were quickly integrated into mainstream jargon yet they somehow failed to produce substantially different outcomes. In this case it is precisely the examination of various institutionally embedded micro-powers that explain the extent to which the dominant discourse of ecological modernization affected the actual regulation of acid rain. Whereas the British case showed many of the problems with the structuration of eco-modernist discourse, the Dutch case illustrates the problems that occur in the phase of discourse institutionalization.

5.2. DUTCH ENVIRONMENTAL POLITICS 1971–1989[1]

One of the key words of Dutch environmental politics is undoubtedly accommodation. Right from the start the discussion of the troublesome relationship between economic growth and environmental quality was not limited to the circles of the counterculture. Indeed, in the early 1970s the legitimacy of industrial society and its practices were in fact a perfectly civilized topic for discussion. In this the apocalyptic message of *Limits to Growth* was the great catalyst. In the spring of 1973 Queen Juliana invited a large number of prominent figures from politics, industry, and academia to discuss the repercussions of the Report to the Club of Rome at the Royal Palace in Amsterdam.[2] The new concerns also solidified in the official assignment of a sixth goal to government economic policy: *leefbaarheid* or quality of life.[3] In October 1971 the Government had created a new Department for *Volksgezondheid en Milieuhygiëne* (Public Health and Environmental Hygiene, VoMil). The new Minister, Stuyt, soon published the White Paper *Urgentienota Milieuhygiëne* (Urgent Memorandum on Environmental Hygiene) which listed the main areas of concern and announced new laws to regulate the problem.[4]

[1] The history of Dutch environmental politics before the 1970s mirrors many of the international trends. As elsewhere, the Netherlands experienced a first wave of environmentalism at the turn of the century (cf. Tellegen 1979). Air pollution first became a real political issue in the Netherlands in the late 1960s, as pollution levels rose and the heightened awareness of environmental risks related pollution to the opaque and élitist decision-making structures. Here it should be remembered that industrialization in the Netherlands did not really gain momentum until after the Second World War. During the post-war reconstruction the Netherlands invested heavily in its industrial base, which, because of the effective reproduction of a social consensus, resulted in a period of unprecedented economic growth. This growth was concentrated in a few sectors, the—heavily polluting—petro-chemical industry being one of the most prominent. In the 1960s the urban smog in the Rhine Delta, the heavily industrialized and densely populated zone around the Rotterdam harbour, was comparable to the situation in the British conurbations and the German Ruhr area. For a good account of Dutch environmentalism see Cramer 1990.
[2] See Stichting Maatschappij en Onderneming 1973.
[3] The other five being a balanced labour market, a stable price-level, a steady balance of payments, a just distribution of income, and stable economic growth.
[4] Most official Parliamentary documents are listed in the references under *Handelingen Tweede Kamer* in order of appearance. Exceptions are the English version of various documents and documents that have been made available through the SDU, the Dutch equivalent of HMSO.

The 1971 *Urgentienota* already revealed some of the particular characteristics of Dutch environmental discourse. Looking at the text one is struck, first of all, by an apocalyptic framing of pollution that very much reflected the argument of *Limits to Growth*. Yet it also indicates a high level of politicization, since the Memorandum does not just deal with the symptoms of decay, but addresses the underlying social practices:

One doubts whether growth is identical to progress now we can see so clearly that the place where growth has to be produced—the environment in the widest sense—indicates its limits ... it becomes increasingly clear that it is the deeply entrenched anti-ecological character of certain forms of human activity in our cultural pattern that is one of the primary reasons for the environmental predicament.[5]

It is essential to appreciate the way in which this structural critique was counterbalanced. A significant feature of this White Paper was that it used apocalyptic pathos in the general framing of the problem of pollution, but was incremental and technocratic in its regulatory ideas. The Urgent Memorandum employed the familiar division of the environment into four compartments: air, water, soil, and organisms and listed mainly local and regional problems that required urgent treatment, one of them being air pollution. What is more, in the outline of the concrete remedial strategies, the idea that the environmental predicament was related to entrenched anti-ecological practices seemed to have been forgotten:

The goal of environmental improvement can be seen as equal to other important societal desiderata among which, next to other goals, also the well known social economic goals. Together this constitutes the total societal goal of welfare in the context of which a continuous political adjustment has to take place regarding the extent to which the various goals can be realised.[6]

Together these statements illuminate the tension between the abstract representation of the environmental problem pregnant with pathos and the practical suggestions to come to a more effective practice of pollution abatement.[7] It is a discursive format that

---

[5] Quoted in Dieleman 1987: 30–1.   [6] Quoted ibid. 32.

[7] Here the Urgent Memorandum was very much unlike the discourse of ecological modernization, in which the paradox between economic growth and environmental care is resolved by drawing on the positive-sum-game format.

combines a moral outcry with a pragmatic orientation to environmental regulation. It also reflects an ambivalent conception of nature. Nature is on the one hand seen as resource ('where growth has to be produced') but its improvement is also a virtue in itself ('a societal desideratum').

The tension between the moral and pragmatic approaches to nature that comes out in the Urgent Memorandum is as obvious as it is understandable given the wider political and economic context of that time. The economic upset of the early 1970s and the world oil crisis of 1973 were accompanied by an oil boycott of the Netherlands.[8] Consequently, the new sixth goal of government policy was almost immediately subordinated to the core concerns of economic performance. Both Hoppe and Honigh have shown that the most positive environmental achievements of the 1970s were in fact the unintended consequence of politically and economically inspired policy, most notably the reduction of the dependency on oil through diversification and energy conservation.[9]

This growing awareness also came to the fore in the context of the parallel discussion on water management. It is important to pay some attention to water management since this issue occupies a unique position in Dutch society.[10] In February 1953 a spring flood combined with high winds inundated large parts of the southeast of the Netherlands and 1,825 people died. A consensus formed that such a disaster should not be allowed to take place again. In 1957 the Dutch Parliament passed the *Deltawet*, which provided for the implementation of the *Deltaplan*, an engineering work of an unprecedented scale, according to which several estuaries would be closed and dikes would be heightened. The Deltaplan gave unprecedented powers to the new *Deltadienst*, the organization that had been erected to enforce the scheme.

The *Deltaplan* has occupied a unique role in Dutch contemporary history ever since its conceptualization. The national trauma meant that the project could count on widespread credibility and legitimacy in the early years. Indeed, it became the central historical

---

[8] During this period the Netherlands was confronted with a complete shutdown of its oil supplies from the OPEC countries because of its friendly relationship with Israel.

[9] Cf. Hoppe 1983; Honigh 1985.

[10] Indeed, the origins of the Dutch political culture of 'consociationalism' are seen as stemming from the medieval struggle against water: see Middendorp 1979. See also Schama 1991.

example of a successful national approach to national crises. The *Deltaplan* is the story-line that embodies the ultimate evidence of the regulatory capability of Dutch government. It gave environmental discourse the historical reference that plays a role very similar to the reference to the London smog and the Clean Air Acts in British air pollution discourse.[11]

The term 'acid rain' entered Dutch public discourse after the 1972 Stockholm conference. It was used to describe additional, far away effects on Scandinavian ecosystems. At first acid rain did not cause either great public or institutional apprehension. Air pollution was primarily perceived as a matter of regulating SO_2-induced urban smogs.[12] Acid rain only gained relevance as the planned extension of the nuclear programme rapidly lost its credibility and the centre-left coalition of the day was, grudgingly, forced to consider reintroducing coal to meet demands. This would result in a rise in SO_2 emission levels even without the predicted rise in demand.

In October 1979 the Government announced the introduction of an SO_2 emission ceiling of 500 m. kg. and came up with the SO_2 *beleidskaderplan* (SO_2 Policy Framework Plan) which outlined how SO_2 emissions were to be contained.[13] Despite the centrality of the traditional concept of a ceiling for pollution, this plan was by no means written according to the traditional understanding of air pollution. In many regards it conceptualized pollution in good eco-modernist fashion. It brought out the SO_2 problem in its full complexity and discussed the transboundary aspects of air pollution. It argued that it would be 'unethical' to be a 'net exporter' of SO_2, signalled the dubious role of tall stacks,

---

[11] Interestingly, even the *Deltaplan* lost much of its numinous legitimacy in the face of the general rise in popular concern over environmental risks. In December 1970 several organizations joined hands in the Comité Samenwerking Oosterschelde Open. They were critical of the great environmental impact of the estuary closure and called for a reconsideration of the last remaining project, the Oosterscheldedam. The controversy over the Oosterschelde became one of the emblematic issues in environmental discourse of the 1970s. For a detailed account of this remarkable case, see Leemans and Geers 1983.

[12] As former minister Ginjaar put it: 'We were totally pre-occupied with the issue of SO_2 emissions.' (Dr L. Ginjaar, Minister for VoMil 1977–1981, interview).

[13] Department van VoMil, 1979. The emission ceiling could hardly be considered as an abatement policy since the actual SO_2 emissions at the time were about 400 m. kg.: see Departement van VROM 1984b: 10.

the effects on human health, and the repercussions for the natural environment:

> [the] harmonious adaptation of human activities must be considered as absolutely essential, in view of the fact that man is now in principle capable of largely annihilating within a few centuries what evolution took thousands of millions of years to produce. From the ethical point alone, man does not have the right to bring close to extermination other types of organisms and ecosystems built on these organisms.[14]

The plan took as starting-points the 'positive duty to maintain the natural environment' as well as the need to curb pollution; it argued for abatement at source, reinforced the polluter pays principle, and called for a balancing of interests and international solidarity. What is more, the SO_2 Policy Framework Plan emphasized the government's conviction that it was possible to reconcile economy and environment. In the Urgent Memorandum the relationship between growth and environment had still been an unresolved issue. In the SO_2 Policy Framework Plan the cost-increasing aspects of specific measures were not the focus of attention, but rather the potential for innovation, for export, and for environmental policies which actually cut costs (to restore corrosion (on buildings), or because of reduced absenteeism (due to health problems), etc.).

Hence the SO_2 Policy Framework Plan indicated that the government was quick to integrate the ideas on environmental pollution that were coming out of the OECD at that time.[15] Most of the emission reduction would have to come from a lowering of the sulphur content of fuels, supported by an energy savings programme to curtail the imminent rise in SO_2 emissions. In addition, the 1979 SO_2 Policy Framework Plan announced experimental projects with fluidized-bed combustion and gasification of coal. The government also stated that the installation of FGD on two 300 MWe power-stations was imminent.

To be sure, the technical debate on SO_2 did not generate widespread public interest. As usual, environmental discourse was much

---

[14] Departement van VoMil 1979: 7.

[15] Departement van VoMil 1979: 90 ff. Reports like *The World Conservation Strategy* (Anon., 1980) or *The 2000 Report to the President* (Anon., 1982) did not have much impact on the Dutch environmental policy domain. Only the *(Voorlopige) Centrale Raad voor de Milieuhygiëne* (CRMH), the Government's official advisory body, published a report on the WCS (see CRMH 1981).

more concerned with issues of high symbolic significance, such as nuclear power and the fate of the seals in the northern Waddenzee. SO_2 abatement remained by and large a technical matter between a few politicians, experts, and policy-makers who were closely monitored by the NGO Natuur & Milieu. Its critique was mostly confined to the fact that it thought the annual emission targets were too high. Furthermore, they contested that these targets were primarily met through the employment of more natural gas, while the refinery industry—a major SO_2 emitter in the Netherlands—was allowed to continue to use its tall stacks. In the face of a major economic recession, however, their complaints did not make much headway at that time. In actual fact it was not until the acid-rain issue restructured air pollution discourse that these matters generated public interest.

One issue managed to revive interest in environmental matters despite the unfavourable economic situation. In 1980 Dutch environmental discourse also received its own Love Canal reference, as public hysteria followed the discovery of large quantities of chemical waste under a number of owner-occupied housing sites that had been constructed in the 1970s. The image of drums of chemical waste being dug out from underneath comfortable terraced houses and of men walking around playgrounds dressed in safety suits wearing gloves and respirators clearly captured the public imagination. The incidents helped the layman understand the problems of a risk society. It illustrated the extent to which the environmental problem, quite literally, had become a structural feature. A major clean-up was called for. Quite suddenly environment specialists inside and outside parliament had the respectability and authority to put these incidents in the appropriate context.[16] Not surprisingly, many other similar cases of soil pollution quickly came to the fore.

While the ground had thus been prepared, acid rain added a whole new dimension to the claim that the environmental problem, requried a structurally different approach. What had started with the local and repulsively concrete and visible problem of soil pollution, now acquired a creeping and invisible counterpart. The issue acid rain was imported from Germany. The international

---

[16] Among themselves they spoke of the 'Lekkerkerk-effect' after the town where the first big waste dump was discovered (Dr R. de Boois, interview).

concern over *Waldsterben* in Central Europe (Germany, Switzerland, and Czechoslovakia) quickly led to the question of whether acidification had a domestic component. In fact there was no immediate evidence of large-scale forest dieback, but there were prolonged rumours about the effects of acidification in confined natural areas with unfavourable soil conditions.

The government's response came immediately after the Stockholm conference of June 1982. Mrs Ineke Lambers-Hacquebard, Junior Minister for Environmental Hygiene, impressed by the reports on acid rain from the Swedish and German representatives, reported that the tall stacks policy was outdated. On a technical level this resulted in the decision to apply FGD scrubbers, to use low sulphur coal, and to improve combustion techniques.[17] The Minister also announced that the Dutch policies towards SO_2 had received widespread support at Stockholm and that the Netherlands was to continue to pioneer the field.[18]

Influenced by the confrontation with the potential scope of the environmental problem, the Government's position started to shift. Although the economic recession was the number one issue, the Government drew on the rhetoric of ecological modernization to incorporate its commitment to environmental care. In the presentation of the budget in 1982 the Government reiterated its argument that in the long run society would benefit from environmental protection and supported its claim with cost–benefit analysis. The rather short-lived centre-left Van Agt Cabinet commissioned various working groups to see which changes in institutional practices were required.

In November 1982 the context of environmental politics changed considerably, as the centre-right Cabinet led by Lubbers came into office. It aimed to roll back the state and sought to organize government policy on a neo-liberal footing. Environment now became the concern of a newly created ministry of VROM, with a responsibility for housing and physical planning as well as for

---

[17] In September 1982 Minister Lambers-Hacquebard announced her intention to have coal-fired power-plants to install, in principle, FGD equipment so as to clean 50% of their flue gases. Old plants were to be retrofitted with two exceptions. By focusing on the electricity industry the government made it possible to exempt industry from installing FGD until 1990, although it had to start using low-sulphur coal. To cut the NO_x emissions the government ordered the usage of low NO_x burners. Together these measures would bring down the SO_2 emissions by 50% and the NO_x emission by 15%.

[18] See CRMH 1983: 136.

environmental management. As part of the neo-liberal turn, one of the major goals of the Lubbers government was to deregulate government policy: to simplify legal rules and procedures, to reduce the number of advisory councils, to cut the permit requirements for business and industry, etc. Indeed, Dr Pieter Winsemius (VVD), the new minister, had been approached to take up the position at VROM because of his work on deregulation for the employers federation VNO.[19]

The change of Government did not bring the change in policy discourse to a halt. In fact the eco-modernist ideas about the deficiency of the legal regulatory approach and the advocacy of ideas about internalizing environmental care had a strong affinity with the neo-liberal goal of deregulation. Under Winsemius ecological modernization became hegemonic. Ecological modernization was the only credible discourse in policy debates (discourse structuration), and the Department also actively started to restructure its organizational practices (discourse institutionalization).

Characteristic of environmental discourse in this period was the interaction between, on the one hand, concern over concrete environmental issues like acid rain, and, on the other hand, the restructuring of the concepts and practices of policy-making. In March 1983 the *Tweede Kamer* (second chamber) of parliament accepted a motion proposed by Dr Rie de Boois, MP of the opposition Labour Party. It expressed concern about the reports on soil acidification, called for an inventory of research into the effects of acid deposition on the soil in the Netherlands, and argued for the preparation of a programme of abatement. In September 1983 the Government announced that the problem of acidification called for a reduction of emissions by a factor of 3 to 4.[20] It argued that its goal would be to prevent the worst damage.

In January 1984 Winsemius and his colleague Gerrit Braks (Senior Minister for Agriculture and Fisheries) presented the government white paper *De problematiek van de Verzuring* (The Problematique of Acidification), the Government's reply to the motion by De Boois of February 1983.[21] An attempt to quantify

---

[19] For deregulation and Dutch environmental policy, see also Hanf 1989.

[20] *Handelingen Tweede Kamer*, 1983–4, 18 100.

[21] The publication included the scientific report *Verzuring door atmosferische depositie* or IWACO-rapport (Departement van VROM 1984a) that was commissioned following the motion from De Boois. The most comprehensive study of Dutch acid-rain policy is Van der Straaten 1990.

the damage estimated an annual cost of 150 to 500 m. Guilders rising to 10 bn. Guilders if policies were not changed. It called for more research into the phenomenon of acidification, but argued at the same time that the remaining uncertainties should not be used to delay emission reductions. Perhaps most significant was the fact that the government indicated that, though the international debate suggested otherwise, acidification in the Netherlands was not simply a matter of sulphur dioxide. Nitrogen (produced by cars and the animal slurry of the infamous Dutch agro-industry) seemed to be at least as important. Despite all this, the Government asserted that a reduction of NO_x by more that a factor of 2 to 2.5 was 'technically not feasible' and that the reduction of NH_3 by more than a factor 2 was 'unrealistic'. For SO_2 the possibilities were suggested to be unclear. However, as far as concrete measures were concerned, the Government proposed to follow the German Large Combustion Plant Directive to combat SO_2 of 1983 and to reduce the NO_x emissions by targeting traffic, which was responsible for about 50 per cent of the emissions. It announced its intention to phase in catalytic converters and lead-free petrol (before October 1986) and suggested that the problem of acidification provided a further reason to stick to the existing 100 km speed limit for cars. For the reduction of NH_3, released through animal slurry, concrete measures were postponed until early 1985. A few months later, in September 1984, the Cabinet announced its intention to retrofit three coal-fired power-stations with FGD equipment at a total cost of 600 m. Guilders. Other coal-fired power-stations were to use low-sulphur coal.[22]

These actions on the acid-rain front coincided with the presentation of a comprehensive new regulatory regime. It was fully based on eco-modernist principles. The *IMP-Milieubeheer 1985–1989* (or IMP-M), the first indicative five-year programme on environmental management,[23] explicitly followed the ecological

[22] This policy package was first officially announced in September 1984 in the *IMP-air 1985–1989* that accompanied the overall *IMP-environmental management 1985–1989* (Ministerie van VROM, 1984b). The Government set as its goal the reduction of SO_2 emissions to 175,000 ton/year in the year 2000 (as opposed to its goal of 350,000 in 1983), while NO_x now only had to come down to 350,000 (versus the goal of 200,000 in the year before). NH_3 had to be cut by half.

[23] *IMP Environmental Management 1985–1989*. It was the materialization of a change that had been in the making since the *Wet algemene bepalingen milieuhygiëne* (General Environmental Regulation Act) of 1980 which was a first attempt to come to a more integrated system of pollution control.

modernization discourse as promoted by the OECD. Pollution was no longer defined as the—inevitable—negative side-effect of production—a clean environment was to be seen as a precondition for further economic growth:

The challenge confronting environmental policy in 1984 has . . . increased rather than decreased. While the first generation of environmental issues has not yet been solved, a new more complex generation has sprung up. . . . It is important that the environment and the economy not be placed in opposition to one another . . . Improving the environment and strengthening a lasting economic development are closely connected and consistent policy objectives. Provided they are managed correctly, environment and economy can be mutually reinforcing.[24]

Indeed, the White Paper could refer to the new Constitution of 1983, which now included an article that read: 'The concern of the government is directed towards the quality of life of the country and towards the protection and improvement of the environment.'[25]

The government thus lent legitimacy to the general call for attention from the radical environmentalists whilst it simultaneously detached itself from the institutional claims of these NGOs. While the NGOs blamed the market and called for radical state intervention, the government's policy discourse introduced a market-oriented theory of implementation instead. The *IMP Environmental Management 1985–1989* marked a significant conceptual innovation.[26] The White Paper announced that the government was to put more weight on economic instruments relative to the traditional legal order of permits and prohibition. It introduced an issue-oriented and polluter-oriented approach to supplement the simple compartmental division of air, water, and soil. From now on the government sought to change the detrimental social practices, aiming to bring about a mental change within the so-called *doelgroepen* (target groups, such as e.g. households or the utility industry). This so-called *verinnerlijking* or internalization became the linchpin of Dutch ecological modernization. At the same time the government announced that it would take certain key problems such

---

[24] The paper explicitly referred to the OECD Conference on Environment and Economics. Page references are to the English version of the *IMP Environmental Management 1985–1989* (see Ministerie van VROM 1984*b*: 12).

[25] Ministerie van VROM 1984*b*: 9.

[26] As noted above, the process of institutional change really dates back to the *Wet algemene bepalingen milieuhygiëne* (WABM) of 1980.

as 'acidification' rather than 'air' as its starting-point.[27] This also meant a shift from effects to causes: it now explicitly targeted various polluting sectors (such as traffic, agriculture, and industry). The policy set 'effect-oriented targets' that had to be fulfilled within eight to ten years and introduced a 'source-oriented policy' aiming at the prevention of pollution at source.

Verinnerlijking: *The New Partnership*

The *IMP Environmental Management 1985–1989* used the discursive principles of ecological modernization to define the parameters within which solutions were to be found. On the one hand the regulation of pollution needed to become more effective, on the other, it was obvious that the upsurge in attention for the environmental predicament was not allowed to interfere with the government commitment to revitalize the Dutch economy and roll back the state. The new approach can be labelled *verinnerlijking* (internalization) after one of its core ideas, and was epitomized by the Minister of the Environment, Pieter Winsemius.

Although we have seen that the turn to a more integrated approach to environmental management pre-dated the coming of Minister Winsemius, it is clear that he left his mark on the restructuring of Dutch environmental policy discourse. Winsemius, a senior consultant at the Amsterdam branch of McKinsey & Company, had strong preconceived ideas about what was to be done. Winsemius was a firm believer in positive management theory, and applied his own general ideas about strategic management to the reorganization of the environmental domain. He saw as his job changing the well-embedded cognitive structures within the department. This was not an easy task. Within the department the struggle against pollution was perceived as an almost hopeless affair. The ever-growing scale of the task ahead and the limited possibilities for change had given rise to what had become known internally as the *Wet van behoud van ellende* (Law of Continued Gloom). This was itself an indication of the discursive power of

---

[27] It listed six policy themes that would receive special attention: *verzuring, vermesting, verspreiding, verwijdering,* and *verstoring* and *verbetering* (respectively, acidification, discharge of fertilizers, diffusion of environmentally harmful substances, waste disposal, disturbance of the environmental equilibrium, and improvement of the policy instruments).

Dutch NGOs, which had introduced the image to capture the idea of government's failure to achieve structural change. Winsemius, however, demanded positive thinking.

I heard that Law of Continued Gloom three times during my first week and then said 'No more where I am!' It makes you sick, unwell. Real Defeatism. . . . I appear to be a true Maslow follower. I got the ideas from Stanford Business School. The idea that you have two sorts of psychology. On the one hand Freud and Jung. They believe that man is sick. You can only shift as best as one can. It will never be O.K. This is a frequently found approach to the environmental issue. I cannot stand it. It makes me feel awful. . . . The other approach argues that man is healthy and searches for excellence. Sometimes it all goes wrong. In that case you help a person. If that does not work you kick them, help them drastically. I was attracted to this idea. You get much more acclaim with it. . . . So my approach is to argue 'Listen it is important that we all act.' This is still possible. You can do an awful lot and it does not make sense to wait for others. Just start. . . . Maybe it sounds somewhat simplistic but it works. People get convinced.[28]

Here we have an example of a story-line that was to function as a discursive crow-bar opening up established commitments. Winsemius simply did not accept the structuralist interpretation of the developments in the environmental domain. Instead he modelled reality according to his own individualist cognitive commitments. Winsemius introduced books like the bestseller by Peters and Waterman, *In Search of Excellence*[29] to the department. The general idea of Peters and Waterman is that it is not the discovery of something that is decisive for innovation, but the implementation. Accordingly, Winsemius argued that VROM should change its focus from the setting of standards to the achievement and enforcement of goals.

Here VROM drew on the image of regulation as a chain of regulatory stages.[30] Whereas until then the emphasis had been on the first three stages of legislation, norm-setting, and the granting of permits, Peters and Waterman would emphasize the importance of the final two: implementation and enforcement. Until then the

---

[28] Dr P. Winsemius, interview.

[29] Peters and Waterman 1982. One of the top civil servants Dr B. C. J. Zoeteman, later combined this outlook with holist ideas in which the world was understood according to Lovelock's metaphor of Gaia: see Zoeteman 1989.

[30] Cf. Winsemius 1986: 79 ff.

guarantee of consensus under government agencies had been the primary concern. Now the aim became to include societal actors as well. Consequently, Winsemius aimed to create a much broader political project through the construction of a society-wide sense of a common purpose.

If the positive approach to environmental regulation was to be successful, motivated subjects were a pre-condition. For this purpose VROM in fact sought to create a new discourse-coalition. First of all by discursively creating a break with the past. The traditional permit practice was retrospectively labelled 'the hierarchical system of command and control' while minister Winsemius emphasized the importance of a new 'partnership' both with other authorities (at the international and national level) and with the target groups in society. He described his *Milieuvernieuwing* (environmental renewal), as the common task for business, households, and government to safeguard the physical conditions for a good environment.

Yet the Department also tried to create a shared view on the reality of environmental politics through the invention of new story-lines on environmental damage and on opportunities for business. Winsemius wanted to further the image that the environmental issue was understandable and thus manageable, and had to be cautious not to build an over-complicated construction of reality into his organizational structure. Consequently Winsemius accepted only five themes and seven to eight target groups.

I simply asked the department to hand in its priorities. Fifty six we got. I said I wanted no more than seven, preferably five. That afternoon we started to sort out the most important. We created a few new words (*verstoring, vermesting*[31]) and suddenly people started to see the inter-relations. . . . And all words had to start with '*ver*': to indicate too much, too fat, too far.

Of course, this represented a great reduction in the actual physical and social complexity of the issue. But this was precisely the point of setting priorities. VROM also defined its target groups as waste-disposal facilities, the chemical industry, the utility industry, the bio-industry, households, refineries, and traffic. Policy-making thus

---

[31] *Vermesting* replaced the obvious technical term 'eutrification'. Winsemius said: 'Eutrification, that word is ridiculous! Nobody will understand what you mean.' (interview).

became a matter of supporting five issues through what became known as the 'policy life cycle' consisting of recognition, policy formulation, solution, and maintenance.[32]

It is an essential feature of Dutch ecological modernization that Winsemius argued that VROM had to learn to take account of the interests of industry. Yet more fundamental seems to have been the actual effort to make individual actors redefine their interests in terms of the discourse of ecological modernization: industry, households, and government agencies alike. Here VROM could draw on the carefully created idea that the environment was on the verge of a breakdown, and that the management of the environmental crisis might, within a couple of years, have to become an almost military operation (see below). Winsemius, however, did not call for a war on environmental decline but suggested a much more moderate regulatory regime. The environment was a serious issue but was a problem that could still be overcome.[33] If one shared the created image of damage, then most certainly *verinnerlijking* seemed a fair deal. For VROM, on the other hand, *verinnerlijking* must also have been appealing. It was hardly the kind of deregulatory move that the Lubbers government had initially announced. If anything, it was much more a re-regulation, with the prospect of a scaling-up of environmental policy-making from a sector-oriented policy to a central policy of government.

In short, the Dutch approach of *verinnerlijking* had complex origins. It was, first of all, an adaptation of the changing ideology of state–society relationships, but it was also an expression of the eco-modernist story-line that environment and economic development were, provided that the process was well managed, not mutually exclusive.[34] A third source was Winsemius's personal commitment to positive management. All three set their own implicit markers on the meaning of ecological modernization.

Furthermore it is obvious that in many respects the *verinnerlijking* approach could thrive on the existing accommodative tradition in Dutch policy-making. Winsemius was not a radical who went

---

[32] Winsemius 1986: 17 ff.

[33] This was another repercussion of Winsemius's positive management theory: 'A losing side does not have supporters. Always make sure people get the idea they are joining a winning team.' (Dr P. Winsemius, interview).

[34] In this respect the *IMP-milieubeheer* frequently referred to the ideas of the OECD conference on Environment and Economy of June 1984.

against the prevailing social bias. In many respects he made changes that reinforced the historical bias towards co-operation between the leaders of politics, government, and business. One could argue that the target-group approach brought social organizations back into the policy-making process, thus breaking away from the 1970s command-and-control model that had been inspired by the American example of environmental regulation.[35]

Verinnerlijking *and Acid Rain*

Acid rain played a key role in this conceptual innovation. First, the new approach had been pioneered in the White Paper *De problematiek van de verzuring* of January 1984. Then acid rain was the first issue to illuminate the limits of conceptual innovation. Despite the innovative rhetoric, the proposed measures to combat acid rain were essentially traditional end-of-pipe solutions, whilst it was already noticeable that the maximum effectiveness of end-of-pipe solutions would be nullified by the sheer growth of volumes (of traffic, cattle stock, and energy consumption). Worse still, scientific findings made it increasingly obvious that the acid-rain problem was rapidly becoming more complex and also related to photo-chemical air pollution (such as that resulting from nitrogen oxide reactions) and Volatile Organic Compounds (VOCs).

This discrepancy did not go unnoticed in the Dutch environmental domain. First, NGOs criticized what they saw as the lenient action by the government on the acid-rain front. Their claim was supported by the Centrale Raad voor de Milieuhygiëne (CRMH), the official advisory body of the Government.[36] Cross-media pollution and the sheer growth in volume constantly annulled the results of governmental action: the burning of sewage to fight soil pollution contributed to the problem of acidification

---

[35] For a discussion of the implicit limits, see Sect. 5.6.

[36] The CRMH emanated from a merger of two advisory councils in May 1981. The CRMH is best understood as the Dutch equivalent of the British Royal Commission on Environmental Pollution. Yet its organizational structure differs markedly from the Royal Commission. The Royal Commission consisted of specialists who were appointed on a personal basis. The CRMH, on the other hand, basically followed the model of the *Sociaal-economische Raad* (SER), the influential tripartite council that formed a cornerstone in the Dutch corporatist model. The CRMH had 25 to 35 members including representatives from business, NGOs, regional and local government bodies, as well as environmental experts. The parity of representatives for workers and employers was required by law.

and the greenhouse effect; the rapid growth in the numbers of cars nullified the effects of the introduction of catalytic converters; and so on. At the same time, many of the intended measures had, in fact, not been implemented in order to avoid conflicts with the government's economic priorities. In March 1985 the NGOs organized a big 'acid-rain week'. It was especially aimed at the Netherlands' biggest acidifier, the Shell oil refineries which at that time were responsible for 30 per cent of SO_2 emissions, but were only marginally affected by the priority programme.

The Turning of the Tide

In July 1986, Ed Nijpels (VVD), the new minister for VROM in the second Lubbers Cabinet, thus inherited a mixed bag from his predecessor Winsemius. The Cabinet was determined to continue its neo-liberal policies. Yet at the same time the failure to meet environmental objectives was becoming apparent. Then the frequently reiterated complaints of NGOs found a more authoritative voice in August 1987, as the *Additioneel verzuringsonderzoek* (Dutch priority programme on acidification, AVO), the official research project into acidification, published its interim evaluation. This research project indicated that the effects of acidification were worse than previously thought, 'An ever ongoing deterioration in species number and vitality can be observed in nearly all forests, heaths and ponds, as well as in other more or less natural eco-systems.'[37] The AVO research indicated that more than 50 per cent of Dutch forest could no longer be categorized as 'vital' and most Dutch ponds were now totally acidified. The study revealed that the acidification of the environment would continue even if the Government's deposition target were to be achieved. It argued that the environment was in fact unable to cope with more than the natural deposition levels (i.e. less than one-third of the Government target load). The report concluded that, if policies were to remain the same, within a few decades the effects of acidification would be worse. This would go against the Government commitment that the worst effects of acidification were to be prevented (i.e. the dying of forests).

Clearly, if Nijpels wanted to get acid rain under control he

[37] Schneider and Bresser 1987: 8.

needed to strengthen the clean-up programme and had to accelerate the pace of implementation. At the same time new emblematic issues like the greenhouse effect and the diminishing ozone layer started to call for attention. Internationally the Brundtland Report added its weight to the legitimacy of the call for action. Yet, though the Cabinet had rather smoothly integrated the message of the Brundtland Commission into its repertoire of story-lines in November 1987, it insisted that a tighter regulatory regime would not be feasible.[38]

In 1987 the latent tension in the Dutch project of ecological modernization became manifest. So far ecological modernization had primarily focused on avoiding cross-media pollution. Considerably less progress had been made in changing policies in other governmental sectors, most notably transport and agriculture. Despite noble intentions, environmental policy by and large remained a sectoral policy that stood next to various other Government commitments. Although specialists, both from NGOs and from the scientific community, had been warning that this strategy was bound to run into difficulties, until then the government had tried to reconcile achieving environmental goals with steady growth in other sectors. In December 1987 these contradictions emerged in the context of an issue with great symbolic appeal. On 4 December 1987 Neelie Smit-Kroes, Senior Minister for Traffic & Water Management, announced the Government's intention to increase the speed-limit from 100 km/hour to 120 km/hour.[39] The new speed-limit was put forward as an element of a sustainable strategy, although a car emits on average 40 per cent more NO_x at 120 km/hour than at 100 km/hour.[40] The announcement generated strong objections from many different quarters.

Moreover, on 23 December 1987 Mr Ed Nijpels, Secretary of

---

[38] See Ministerie van VROM 1987*b*. From that time the quotation of the definition of sustainable development would feature in all relevant policy documents. For the feasibility argument, see Ministerie van VROM 1987*a*.

[39] The speed-limit was changed to 120 km/hour on 83% of the motorways. On the remaining 17%—the busiest parts, mainly in the western Randstad—the speed-limit would remain at 100 km/hour mainly 'for safety reasons.' See *Handelingen Tweede Kamer*, 1987–8, 20 366, no. 1–2.

[40] The argument was that nobody complied with the 100 km/hour, which made it impossible to enforce the speed-limit effectively. Most people drive at 120 km/hour. If a speed-limit of 120 km/hour could be enforced, the average speed would go down and the NO_x emissions would come down proportionally. This would imply a 5% reduction of NO_x emissions.

State for VROM, published the *Tussentijdse evaluatie verzurings-beleid* (Interim Evaluation of Acidification Policy).[41] The Memorandum recognized that the Government targets of 1984 had not been met. Statistics showed that the main reason was growth in volume, in energy consumption, traffic, and agriculture.[42] The Government argued that a new policy would require a reduction in deposition levels of 60 to 90 per cent but did not introduce new policy targets. The Government announced that it would wait for the final results of the Priority Programme before outlining new policies. In the mean time the Government announced that it intended to 'exchange ideas' with various social and business groups to create an optimal social basis for a more stringent policy.

The Return of the Apocalypse

In December 1988 the RIVM, the leading research institute on public health and the environment, published the scientific report *Zorgen voor morgen* ('In Care of Tomorrow').[43] It sketched a dramatic picture in which the environmental crisis was not presented in terms of a small number of well-known issues such as acid rain or the diminishing ozone layer. The environmental problematique was depicted as a structural crisis. The environmental problematique ranged from global threats like climate change and rising sea-levels to the local pollution of soil and drinking-water reserves. As far as acid rain was concerned *Zorgen voor morgen* showed that even if all technical measures were to be taken (FGD retrofits, catalytic converters, sludge recycling) the Government goals of prevention of worst damage would not be achieved.[44] In his foreword minister Nijpels argued

---

[41] Ministerie van VROM 1987*b*. The Memorandum was a joint product of VROM together with the Ministries of Traffic and Water Management and Agriculture and Fisheries.

[42] The intended SO_2 reduction in the IMP air 1985–1989 was 70%, but only 50% had been achieved, NO_x 33% had grown 4% and NH_3 should come down 50% but only 25% was achieved. All these increases were caused by a greater economic growth than expected. In the case of ammonia there was also implementation failure (Ministerie van VROM 1987*a*: 3).

[43] Langeweg 1988.

[44] SO_2 should go down by 90%, NO_x by 70%, and NH_3 by 80%. These figures as such were not new to specialists. The importance of *Zorgen voor Morgen* was the presentation of these figures to a new and wider audience by an institute that had the ethos of respectability and objectivity.

This survey forces a contemplation on the way in which our society meets its wants . . . Taking this report into consideration I am of the opinion that far reaching readjustments of our relation to the environment should not be postponed, even though we have to account for many uncertainties.[45]

This apocalyptic interpretation of the state of the environment was boosted when, in the same month, Queen Beatrix devoted her 1988 Christmas address entirely to the environmental crisis and the future of mankind. She warned 'slowly the earth is dying and the inconceivable—the end of life itself—becomes thinkable.'[46] The message was the exact opposite of the words the Cabinet had put in her mouth at the opening address of the Parliamentary year in September 1988. The latter address had included a passage in which the Government argued that the quality of air and water had improved. Hence, giving the opposite view in her Christmas address, the Queen not only added her influence to the critical interpretation of the state of the environment, but she also implicitly accused the Cabinet of irresponsibility. The effect of this Royal and scientific blessing was that Dutch ecological modernization, which had until then been a matter of target loads, critical limits, and effectiveness and feasibility studies, now became a matter of integrity and civilization.

December 1987 saw the widespread resurgence of a pathos-laden apocalyptic environmentalism that reminded many commentators of the Dutch reception for *Limits to Growth* in 1972.[47] In January 1989 employers organizations and trade unions announced that they would cooperate in the fight against environmental pollution, the social democratic Partij van de Arbeid argued it would give the environment priority over individual spending-power, while minister Nijpels underlined the new sense of consensus on the need for a national approach to the environment as he exclaimed 'if only we had listened earlier to these dyed-in-the-wool Greens.'

The new consensus seemed to pave the way for the introduction of a comprehensive plan of action to fight the national environmental crisis, a *Deltaplan* for the environment. This was exactly what was in the making. The Ministry of VROM was preparing its first *Nationaal milieubeleidsplan* (National Environmental Policy

[45] E. H. T. M. Nijpels, in Langeweg 1988: viii.
[46] Christmas address as printed in *NRC Handelsblad*, 27 Dec. 1988.
[47] Caljé 1989.

Plan, NMP). After tough fights within the Cabinet, Minister Nijpels presented his long awaited *Nationaal milieubeleidsplan* in May 1989. It came to be the best orchestrated anti-climax of recent Dutch political history.[48]

First, the Cabinet fell over one—minor—element of the NMP package on 2 May 1989, even before the NMP had been formally presented. The Dutch thereby became the first nation to topple a government over an environmental issue. Then on 25 May 1989 the NMP was brought to parliament. Its subtitle *Kiezen of verliezen* (Choose or Lose) suggested that the NMP was the political programme to match the definition of the environmental predicament as described in *Zorgen voor morgen*.[49] The latter report had indicated that 80 to 90 per cent reductions in emissions were necessary to prevent further deterioration of the environment. Yet the NMP did not follow the recommendations of *Zorgen voor morgen*, but suggested 'broadening', 'intensifying', and 'elaborating' existing policies.[50] In fact the NMP presented a policy package that fell seriously short of the official government goals. Once more acid rain became the metaphor through which many understood the NMP. Here 'to choose' would allow for the survival of 20 per cent of Dutch forests: 'An interim goal of 2400 acid equivalents is used for the year 2000. This allows for the protection of 20 per cent of Dutch forests.'[51] By first constructing an apocalyptic image of the environmental crisis and subsequently presenting a policy package that did not match its very own description of the problem, the Government had inflicted the perfect legitimacy crisis on itself.

5.3. ACID RAIN AS TEXT

In many respects the Dutch acid-rain controversy seems the mirror-image of the British controversy. Indeed, especially in the early years it seems questionable whether the word 'controversy' is

[48] See Hajer 1992*a*.
[49] The apocalyptic vision was by no means restricted to the subtitle: see n. 4 of Chap. 1 for comments on the front cover. The section on acidification was illustrated by a photo of a woman dressed in black walking under dead oak trees (*Handelingen Tweede Kamer*, 1988–9, 21 137, No. 1–2: 39).
[50] Ibid. 5, 7.   [51] Ibid. 133.

applicable to the Dutch debate. After all, instead of a long quarrel over the question of whether acid rain was a serious problem, acid rain was in the first instance simply integrated into established concerns over sulphur dioxide pollution. Care for the natural environment was added to the concerns over the urban realm. Furthermore, when acid rain subsequently came to dominate discursive space, everybody joined in a general part-singing of apocalyptic hymns. But this was not simply propelled by an effective NGO lobby, Minister Winsemius himself immediately took to the pulpit. At the key moment in February 1983 when Dr Rie de Boois, MP for the opposition Labour Party, proposed a motion in the Standing Committee on Environmental Protection of the Second Chamber of Parliament, he replied as follows:

Only now we start to realise its magnitude. I do not want to cause panic, but it proves to be a considerable problem . . . It is extremely complicated and we do not know enough. We do know, however, the points of application, traffic, electricity utilities and presumably intensive agriculture. . . . We must not waste time chasing everything. We must get the crocodiles—the big things—out of the water.[52]

Winsemius argued that acidification was one of the major problems of our time, welcomed the motion, and argued that it supported his policy.

Yet this is not to be taken for evidence of the lack of a controversy. On the contrary, it is one of the defining characteristics of the Dutch acid-rain controversy. The principal feature of this controversy was the steady separation of the discourse on the definition of the problem and the damage caused from the discourse on regulatory measures. This distinction existed both in time (with separate meetings discussing damage and discussing measures), as in place (with different groups or councils discussing either the problem at large or the precise regulatory response). However, the distinction of the two chambers of Dutch environmental discourse should not be reduced to the exercise of power by one specific key actor. In actual fact, it is a structural feature which is well embedded in Dutch history and in the Dutch consciousness. Environmental NGOs only sought to reinforce the apocalyptic discourse by pointing out that abatement strategies

---

[52] *Handelingen Tweede Kamer*, 1982–3*b*, UCV 17, 21 Feb. 1983, 43.

had to realize a 75 per cent reduction within ten years, otherwise the European forests would be 'lost forever'.[53]

As the Standing Committee on the Environment discussed the White Paper *De problematiek van de verzuring* (the Government's response to the De Boois motion) in May 1984, political actors once more reproduced the apocalyptic framing of the acid-rain problem. Both the Christian Democrats and the liberal democrats of D'66 were quick to position acid rain in the context of the crisis of industrial society:

With acidification our society presently experiences the dark side of our welfare. Still, for people working within the environmental sphere acidi- fication has a positive side, be it a bitter one. The public attention to the environment has increased and now air pollution is shown to have visible effects, the dying of forests, policy measures are more acceptable for society.[54]

Acid rain was positioned as a stepping-stone to a new approach. As Dick Tommel, MP for D'66, said in May 1984: 'We are now being presented the bill for chasing after material welfare for decades, disregarding the negative effects. We have been spending the pool at the cost of the environment.'[55] All MPs agreed that 'acid deposition (was) one of the most serious environmental prob- lems of the decade'.

Apocalyptic rhetoric clearly played an instrumental role in the social accommodation of the—latent—conflict over acid rain. The Government suggested that it recognized that serious action was required, but more importantly it seized the opportunity of taking the initiative away from its critics. The part-singing took the issue off the streets and relegated the issue, for the time being, to the sphere of experts and policy-makers.

In many respects the linchpin of the Dutch controversy was the question of what constituted the appropriate form of regulation. The credible answer to the question came from an expert study that was conducted for the Department of VROM. It came out strongly in favour of targeting the abatement of SO_2, since this

---

[53] See Van Ooyen and De la Court 1984: 67.
[54] Mrs Oomen-Ruijten (CDA), in *Handelingen Tweede Kamer*, 1983–4, UCV 95, 14 May 1984, 1.
[55] Mr Tommel (D'66), in *Handelingen Tweede Kamer*, 1983–4, UCV 95, 14 May 1984, 14.

would be more cost-effective than policies aimed at curbing NO_x or NH_3 pollution.[56] Now, what seems to capture the contradictory nature of Dutch environmental discourse is the complete mismatch between the society-wide apocalyptic framing of the problem and the acceptance in parliament of what was in fact a rather modest policy package. Insisting on the apocalyptic definition of the problem, NGOs could easily reject the decision to focus on just one pollutant instead of attacking the issue at large and denounce the package as 'too little'. Yet NGOs now seemed a voice in the wilderness. The very same MPs who had expressed their extreme concern over the acid-rain problem, now left their pity at home and by and large quickly and rather uncritically endorsed the modest policy measures.

So the Dutch debate on acid rain can be aptly described as a 'controversy'. Yet whereas the British controversy was rather clear-cut, one would be hard pushed to distinguish two really distinct story-lines on the basis of the preliminary account given above. In fact all the main actors within the argumentative game used apocalyptic rhetoric in their construction of the problem and, likewise, they all drew on eco-modernist concepts and constructs in the expression of their ideas of the appropriate regulatory response.

This difference in the way in which the acid-rain controversy was conducted in Britain and the Netherlands reflects well-embedded cultural differences between the two countries. British debates have strong antagonistic features. Positioning is not only a matter of having a persuasive argument but also of being able to show the inferiority of the argument of 'the other side'. In the Dutch context debates have strong inclusionary features. Positioning is mostly characterized by showing that an argument is in fact to the benefit of all. Yet both discursive systems have strong rules of prohibition, of exclusion, and of discipline. These rules, and the discursive order that together they sustain, can be illuminated in our analysis of the various practices in the Dutch acid-rain controversy. So, while it would be rather hard to distinguish two distinct story-lines in advance, it might well be that the analysis of practical behaviour allows us to reconstruct two varieties within Dutch eco-modernist discourse.

---

[56] See Tangena, 1984. Interestingly, the study left more structural measures, such as energy conservation, undiscussed.

5.4. THE CONSTRUCTION OF DAMAGE IN THE NETHERLANDS: THE POWER OF APOCALYPSE

> Acidification! . . . We are flooded by reports on the damaging effects of acidification. Reports on the thousands of dead lakes in Scandinavia and Canada; Reports on dying forests in West and Central Europe . . . Reports on acidification of soil, ground water and surface water and the harmful effects of all this for the provision of drinking-water; and finally reports on the damage to materials with as most persuasive examples the damage to historic monuments, sculpture and stained glass window panels . . . the problem of acidification is to be taken very seriously indeed.
>
> Minister Pieter Winsemius, 1984[57]

In 1972 the image of air pollution damage reflected a bias to urban smogs similar to those found in Britain at that time. The focus of Dutch air-pollution control was on ambient SO_2 levels and health concerns. Yet the practices through which these images were reproduced were different in character. In the Netherlands air-pollution control was not an insulated technical sphere governed by experts. Rather it was a political activity that was closely monitored by (mostly social-democratic) politicians who often saw air pollution as a campaigning issue. It is not surprising, therefore, that the critical ideas that dominated among academics at that time, found their way into the practice of air-pollution control. Perhaps only few thought that air pollution was to be understood directly in terms of class struggle, but a much larger group saw air pollution as a matter of injustice. As one air-pollution inspector at that time argued

> We were really searching for arguments just to be able to condemn air pollution and those who caused it. Together with scientists, yes, it was a, you could say tendentious way of seeing things. . . . Behind it was the 1960s idea that we had a mission. Industry was bad. You really were not allowed to work for industry. Profit was dirty, it was polluting. It really was *The Way We Were*. Do you know that film?[58]

---

[57] Opening speech at the Symposium on acid rain held at Den Bosch, 17–18 Nov. 1983: see Adema and Van Ham 1984: 17.
[58] S. Zwerver, interview.

Although one official cannot epitomize a whole group, his description certainly matches a broader characteristic of Dutch air-pollution regulation: a high degree of politicization and interaction between regulation and society. Consequently, air-pollution discourse was not a consultative debate, it was structured around the distinction of polluters (industry) and polluted (health effects). Later the distinction between polluters and polluted also predetermined the way in which the reports on the negative effects of air pollution on nature were received: not only did industrial emissions affect health, they also damaged the natural environment.

Yet interestingly, the introduction of acid rain in the environmental debates did not, at first, make the Dutch aware of domestic damage. The publication of the OECD report on long-range transport of air pollution in 1977 created a foreign image of damage through the recognition of the existence of acid rain in Sweden.[59] As Dr Leendert Ginjaar, Minister for VoMil from 1977 to 1981, put it 'In retrospect it is a strange idea that we all spoke of acid rain as a problem relating to the Swedish lakes but that nobody was aware that if these lakes acidified, soil might be acidifying too. It was a blind spot.'[60] The uncritical import of the Scandinavian acid-rain story not only bemused the cognition of politicians but also of scientists:

the label acid rain has, in fact, forced research in a particular direction. In actuality the rain in the Netherlands in not that acid. Yet this is precisely because of the role of ammonia. But it was not until Van Aalst started to work on dry deposition [in 1982] that we came to realise that dry deposition was much more important than actual rain.[61]

Hence in the Netherlands the import of the 'acid-rain' story at first only reinforced the bias towards sulphur dioxide. The persistence of this bias was, at first, not the result of active agency but of the prevailing cognitive structures that hindered the perception of the acidification problem in full. The common-sense idea that ammonia is basic, not acid, inhibited the acceptance of ammonia as acidifier. It led to a lot of letter writing to newspapers, and

[59] OECD 1977.

[60] Dr L. Ginjaar, interview. In actual fact the Rijksinstituut voor Natuurbeheer (RIN) had reported on acidification in the Netherlands since the early 1970s. Yet they were not taken seriously until acid rain became a national political issue in 1982.

[61] Professor N. van Breemen, interview.

even at the ministerial level common sense prevailed. Dr Pieter Winsemius, a physicist by training, recalled that

At first I did not understand the involvement of ammonia at all. I literally asked for the complete chemical formulas of acid rain. I got ten formulas with a microbe that did something mysterious somewhere in between. It transformed a base into acid. Hallelujah. Try and explain that to people.[62]

Once the rumours about the implication of ammonia became public the agricultural lobby started to draw on the common-sense perception of the acid-rain problem. Yet more significant was the fact that, like in Britain, the old biases had solidified in the organization of the monitoring network. In the densely populated Netherlands the bias to urban air pollution came out in the fact that the Inspectorate initially only monitored SO_2 depositions.[63] Hence one could not draw on the monitoring results to show correlations between dry deposition and acidification damage. Until 1982, therefore, the Dutch debate was hampered both by inadequate knowledge and inadequate data. So it was the interaction between three factors that initially held up the recognition of the particular Dutch air-pollution problem: (1) the cognitive construction of air pollution as a problem of urban smog, (2) the institutional SO_2 bias of the monitoring network, and (3) the discursive framing of acid rain as a Scandinavian problem.

Only in 1982 did it dawn upon the Dutch policy-makers that air pollution also damaged the Dutch natural environment. Apart from the growing consensus among scientists on the implication of dry deposition in Dutch acid rain (which should therefore better be called 'acidification'), it was the Stockholm Conference that opened the image of damage. The authorized German acid-rain story made it much more likely that the Netherlands' natural environment would be affected too. A paradoxical feature of this suspicion by analogy was that the German reports concerned not

[62] Dr P. Winsemius, interview.
[63] The Air Pollution Act of 1972 gave the government the means to operate one of the most precise SO_2-monitoring networks in Europe. It was initiated by the RIVN in 1966 and since 1976 has operated some 220 stations. The funding for operating the network comes from the charges on fuel that were contained in the Air Pollution Act of 1972. Apart from this grid of monitoring stations, the Netherlands has 120 additional stations in urban areas and on the borders. As far as other pollutants are concerned, the Netherlands has 92 monitoring stations for NO and 30 for ozone. See *Handelingen Tweede Kamer* 1982–3, IMP-Lucht 1981–1985, 17 600 XVII no. 7.

tree dieback but the sudden death of forests. But until that time nothing even vaguely similar had been reported in the Netherlands. Still the German story resulted in an argumentative game in which a Dutch image of damage was (socially) constructed. Three practices were particularly important here: tree health surveys; excursions and symposia; and public awareness campaigns.

The Sub-politics of Tree Health Surveys

Once acid rain had obtained the central position in air-pollution discourse the interest in forests grew. As journalists started to report on the magnitude of *Das Waldsterben* in West Germany, so Dutch policy-makers, environmentalists, and foresters started to wonder about the vitality of Dutch trees. Environmentalists saw forest damage as their big chance to win credibility. 'Our idea was that if we could educate the people so that they could recognise damage for themselves, they would be frightened to death.'[64] Likewise progressive politicians recognized the potential of the image of dying forests for their struggle against industrial polluters.

I have to admit I was quite happy with the reports in the decline of the Black Forest . . . Dying forests were a symptom that you could use to fight air pollution . . . You needed dying forests to prove it was damaging, that it was not just fantasy. We really played on the sentiments of a deforested Veluwe with hotels and pensions closing down.[65]

It led to the De Boois motion of 21 February 1983 which called for the prepartion of an inventory of the effects of acid deposition in the Netherlands and for the preparation of a programme of abatement. This parliamentary debate triggered new media attention. The new, domestic image of doom reached the nation as the Dutch Television News broadcast an item in which two foresters from Staatsbosbeheer[66] demonstrated the scale of *Waldsterben* in

---

[64] T. de la Court, interview. For this purpose the NGO WISE also published a booklet to help people recognize acid-rain damage: see Distelbrink *et al.* 1985.

[65] Dr R. de Boois, interview. The Veluwe is one of the main natural reserves of the Netherlands, with mainly poor sandy soils. De Boois did indeed use this image of desolated resorts in the parliamentary debates: e.g. *Handelingen Tweede Kamer*, 1983–4, UCV 41, 19 Dec. 1984, 21.

[66] *Staatsbosbeheer* is best understood as the Dutch equivalent to the British Forestry Commission. However, contrary to the UK where the FC and Nature Conservancy Council are separate institutions, *Staatsbosbeheer* combines the production interest with a responsibility for conservation.

Germany.[67] The Germans argued that acid rain was killing their trees but did we know what was going on in the Netherlands? Were the Dutch forests dying or not? Dutch experts were simply unable to come up with an answer. The initial response was to conduct a tree health survey.

Following the De Boois motion, Staatsbosbeheer conducted a pilot survey of tree health in 1983. The report stated: 'the vitality of Dutch forests is alarming. The low vitality cannot be explained on the basis of normal forestry causes such as illness, plagues, groundwater extraction, bad choice of crop or unfortunate origin.'[68] The sub-text was that air pollution was thought to be implicated. In 1984 the first regular survey was conducted. Staatsbosbeheer characterized the situation as alarming or critical. Again air pollution was held responsible.[69] In the subsequent surveys the percentage of non-vital trees steadily increased. In 1987 the percentage of vital trees had decreased to 42.6 per cent. The survey concluded that

If environmental policy fails to bring about sweeping reductions of emission levels of various air-polluting substances, one has to count with the fact that over a period of time air pollution will become all decisive for the vitality of our forests as well as other ecosystems.[70]

The ideas about the causes of the change were constructed around synergetic ideas such as the multiple stress hypothesis: 'Vital forests can reasonably be expected to recover from damage caused by weather circumstances, insects and fungi, etc. The influence of these factors can be aggravated by air pollution.'[71] Likewise, the 1987 survey gave examples of insects that used to be quite harmless which were now thought to cause considerable damage.[72] Nevertheless, there was a structural mismatch between the outcome of the annual surveys next to the public perception of domestic acid-rain-induced damage. The increase in the percentage of 'non-vital' tree was not nearly as dramatic as in Britain. In fact a compilation of survey results seems to indicate stability more than anything else.[73]

[67] Item by Marijn de Koning for *NOS-Journaal*, 25 Feb. 1983, with W. M. J. den Boer and J. C. A. M. Bervaes (*NOS-Journaal*, 1983).
[68] Den Boer and Bastiaans 1984: 71.   [69] Anon. 1984: 19.
[70] Anon. 1987: 13.   [71] Anon. 1986a: 18.   [72] Anon. 1987: 19.
[73] Because of a difference in methodology the results of the 1983 pilot survey could not be integrated.

Survey reports were perhaps the single most powerful practice in the creation of an image of damage. In rhetorical terms it was a practice that scored high both on *ethos* and *logos*. It was through surveys that socially respected and knowledgeable institutes like Staatsbosbeheer presented their expert accounts of the state of the environment. Furthermore, surveys gave a numerical representation of damage and produced a concrete figure to replace the varied mix of rumours, accusations, and incidental reports of damage. What is more, tree health surveys had a high political importance because of the semiotics of a tree. As the quotations presented above already indicate, trees themselves had considerable symbolic importance and were widely used as biological indicators of the morality of modern civilization.

The rhetorical power of tree health surveys was not only recognized by environmentalists and politicians. Policy-makers wanted to stop forest damage becoming a runaway issue. For them surveys were a means to regain control over the issue. For scientists surveys were a way of freeing acid rain from the discursive domination by bad journalism.[74] While society called for action, and politicians called for a definitive scientific assessment, scientists felt increasingly uneasy with the results of the surveys. They became aware that understanding forest dieback was not a matter of filling in the gaps of knowledge. It became apparent that a scientific understanding of the process was actually hard to achieve. So initially researchers concluded that about 80 per cent of the forest was dying, but later this conclusion was withdrawn, not because it was necessarily wrong but simply because the problem was more complicated. This tension between the political quest for numbers and the high levels of indeterminacy in scientific circles provided the basis for all sorts of social forms of manipulation of expert knowledge. A more detailed analysis of the actual practice of representing damage through vitality surveys can provide an insight into this process of social closure and the construction of acid-rain damage.

The representation of forest health in Table 5.1 does not indicate a huge problem and hardly matched any apocalyptic claims or legitimized a crash regulatory programme. Yet the interesting thing about the Dutch tree health surveys is not what finally found

---

[74] G. van Tol and T. F. C. Smits, interview.

TABLE 5.1. *The representation of forest damage in percentages,*
1984–1991

Year	Vital	Less vital	Barely vital	Not vital
1984	50.8	39.7	8.0	1.5
1985	49.9	35.0	13.0	2.1
1986	46.9	32.0	16.0	5.1
1987	42.6	36.1	16.6	4.7
1988	50.9	28.1	16.0	5.0
1989	50.1	30.7	15.8	3.4
1990	52.5	28.7	15.3	3.5
1991	52.0	30.2	14.7	3.1

Source: Smits 1991: 11.
Note: The Dutch categorization is *vitaal, minder vitaal, weinig vitaal, niet vitaal*.

its way into the reports but what was left out. As politicians
needed input, forestry policy-makers were called upon to produce
data. They found, however, that there was no agreement on something as fundamental as what a healthy tree looks like. In this
discursive vacuum the German example at first played an important role 'These German ideas about trees losing their needles were
of tremendous importance. Seeing trees with yellowing needles
and then thinking "well, this must be acid rain"'.[75] Here similarity
became evidence. Yet the scientific policy advisors who compiled
the surveys of course needed a more precise starting-point. They
had to find a way to measure the condition of the forests over
time. The specialized expert knowledge of foresters was not of
much use. As elsewhere, the Dutch forestry research institutes
were predominantly working on issues related to growth and yield,
not on an abstract issue like forest health. Consequently, an image
of a healthy tree was constructed on the basis of a combination
of herbarium material, literature review, old photographs in peri-
odicals and monographs, and various field observations.[76]

---

[75] Dr T. Schneider, interview. The German example was also essential for the
environmental movement. The NGO WISE booklet *Zelf schade herkennen aan
bomen* drew heavily on the German publication *So stirbt der Wald* (Schütst *et al.*
1983). T. de la Court, interview.
[76] Dr W. M. J. den Boer, interview.

To establish the indicators of tree health, Staatsbosbeheer also drew on the advice of the aforementioned FAO/ECE working group on the Effects of Air Pollutants on Forest and the Landesanstalt für Ökologie, Landschaftentwicklung und Forstplanung of North Rhine Westfalia.[77] However, to its dismay Staatsbosbeheer found that if it were to follow the international standard, most of Dutch trees should be classified as 'unhealthy'. It was therefore decided to adopt less rigid standards.[78] Hence although the herbarium material showed, for instance, that a Scots Pine could have four years of needles in the Dutch climate, it was decided to define a healthy tree as having two years of needles. This shows how the image of the so-called tree of reference, the very basis of the survey reports and consequently also of political decision-making, was a politically inspired social construct.[79]

One may play down the importance of the sub-political choice of the tree of reference. After all, the really important thing was to monitor the relative change. So even if one were to start from relatively favourable assumptions, decline would show over the years. Table 5.1 does not indicate any serious decline but did this imply that there was no decline? Here one has to appreciate the influence of two seemingly minor statistical practices. First the statistical practice of *teruglegging* or 'replacement'. This refers to the fact that the statistics of the surveys gave only relative percentages. This implied that each tree that died or was felled disappeared from the statistics the next year. In this way Dutch forests could all but disappear without any indication in the survey statistics. Secondly, the problem of classification and aggregation. The survey distinguishes four classes of tree 'vigorousness'. But which classes could legitimately be added to come to a general conclusion on forest health? What is more, the surveys were aggregated at provincial level. This implied the submergence from the statistical representation of the disappearance of one tree species in one specific region. This representational problem was reinforced by a third forestry practice, so-called 'sanitation felling'. If a forest was well maintained and old or sick trees were cleared, the longitudinal interpretation of the surveys would be rather

---

[77] Anon. 1984: 4.   [78] J. C. A. M. Bervaes, interview.
[79] On the other hand one has to be aware that there can be legitimate reasons to adjust the classification. Certain trees might indeed live on the edge of their territory and therefore always look slightly wretched.

difficult if not impossible. Hence, the Dutch tree health surveys were, in fact, a rather dubious way of representing nature.

It will not come as a surprise that the construction of damage in the tree health surveys hid an intense controversy. The central figure in the tree health controversy was undoubtedly Dr Mauk den Boer, a policy-maker from the forestry section of the Ministry of Agriculture and Fisheries and author of the first vitality surveys.[80] Since the mid-1970s he had been convinced that something was 'terribly wrong' with the Dutch forests. He was interested to see whether one could say something about the extra factor of air pollution and confronted those who tried to understand the new phenomenon in terms of well-known categories such as frost, drought, or insects. As from 1982 Den Boer took the issue to the press and became known as the national expert on forest damage. Initially, Staatsbosbeheer supported his initiatives, yet they gradually changed their attitude as the role of ammonia became more prominent. The potential implications of a clear image of damage that would show how agricultural practices were implicated caused the Ministry of Agriculture and Fisheries to step in and caution Staatsbosbeheer.[81]

Den Boer had identified the tree health surveys as a useful instrument to influence politics. For him the surveys were to be a signal. 'Surveys are of course an extremely primitive method. I saw it as a way to get people to think about the forest. Just give them a sign: two hundred experts have looked at the forests and they all think it is in bad shape . . . I also saw it as a means to get real research going.'[82] Yet Den Boer did a lot more than was actually published in the survey reports. In 1984 he ran SPSS on the survey data. It illuminated a direct relationship between forest health and intensive agricultural practices. 'These relationships [between location of bio-industries, i.e. ammonia emission and tree dieback] were astonishingly clear. It was not the water household, it was not the exposure [to wind] . . . You could relate all kinds of things, wind, water, bad soil: they did not show.' However,

---

[80] He was mentioned as a key actor by all interviewees who had been involved in acid-rain research or policy-making.

[81] In 1990 the Ministry's name changed into *Landbouw, Natuurbeheer en Visserij* (Agriculture, Nature Conservation, and Fishery) to include the responsibility for conservation. However, agriculture remained the main policy concern.

[82] Dr W. M. J. den Boer, interview.

this statistical component was taken out of the draft survey report. Den Boer subsequently tried another representational device: a coloured map. 'Here the correlations really hit you. Wonderful. That map really gave it all away. Especially the relationship with ammonia could not be missed.' Initially the map was not going to be included in the report but by taking the initiative to put the map on the ANP telex[83] Den Boer made it part of the public domain.[84] He thus confronted his superiors with a *fait accompli* and the map was included in the report. The map proved to be a powerful way of communicating the survey results. In the meeting of the parliamentary Standing Committee on the Environment of 10 December 1984 Dr de Boois referred to the pattern of 'yellow and red dots' and argued: 'The Staatsbosbeheer map indicates that [areas of] bio-industry and affected forests to a large degree coincide'.[85] Although Den Boer remained in charge of the actual surveying until 1986, his say in the annual survey reports diminished. Any information that could be to the disadvantage of the department was eliminated.[86] In 1985 Den Boer was no longer allowed to do the presentation of the annual survey. In 1986 Den Boer was transferred from the Ministry to one of the district offices of the Forestry Department.

Den Boer, a classic example of a whistle-blower, opened alternative channels of communication. He not only secured the introduction of his findings into the debate through his links with the media, but also by leaking them to the staff of the rival Ministry of VROM. VROM subsequently used this information in the interdepartmental debates and also commissioned scientists to work on the precise relationships between ammonia levels and forest dieback (see Section 5.5). In all, it shows the degree to which the autonomous action of individual officials can extend the reflexivity of decision-making. Such actions are, of course, unthinkable in political cultures which extensively control the loyalty of officials to official government policy, such as in Britain under the Official

---

[83] ANP is the main Dutch news agency.

[84] This intervention caused him a lot of trouble: 'I was told this was really not on. But I told them, this is objective information and you are trying to prevent me from bringing this out? Once it was on the telex they had to give in.' Dr W. M. J. den Boer, interview.

[85] *Handelingen Tweede Kamer*, 1984–5, UVC 32, 42. The map was also reprinted in the widely read weekly *Vrij Nederland*, 15 Dec. 1984: 7.

[86] Main gatekeeper was the *Directie algemene zaken, milieu en planologie* of the Ministry of L and V.

Secrets Act and the extensive controls that are imposed on civil servants of the European Union.

The Importance of Excursions and Symposia

Tree health surveys were only one element in the construction of the public image of damage. As in Britain excursions were another essential link in the chain of persuasion. Interestingly, here the Department for VROM closely co-operated with NGOs. Jan Fransen, a leading campaigner with Natuur & Milieu, organized many excursions to damaged forests both for journalists and MPs. He argued 'It helped their imagination. Suddenly all the scientific information got a much greater weight than when they had not actually seen the effects. Excursions were essential eye-openers.'[87] Dr Rie de Boois, environment spokesperson for the social-democratic party PvdA and a biologist by training, saw excursions as a way of showing her colleagues the concrete effects of policy-making or the lack thereof. Again it was Dr den Boer who used this practice at a key moment. On the weekend prior to the parliamentary debate on the *Notitie inzake verzuring* Den Boer organized a trip to the German forests at Paderborn with a group of MPs. In the debate nearly all MPs referred to what they saw during the visit. The effects were described as 'disconcerting and bewildering', 'dramatic', and 'grievous'.[88] The discursive entrepreneur was clearly satisfied: 'In parliament one really had this fresh emotion and the firm wish that something should happen. It was perfect timing.'[89]

But obviously, visits are a rather dubious sort of persuasion. Laymen can be shown trees that are dead or are obviously in bad condition. Respondents put it to me that more than once, visits were made to places where a hundred metres down the road trees were flourishing. As a leading scientist recognized 'Basically it is an exercise in rhetoric.'[90] This is even more true for the new generation of environmental problems than for traditional environmental problems. Here the point precisely is that the problems escape sensual perception but there is an undeniable trend to get access to these issues through familiar practices. Be that as it may,

[87] J. Fransen, interview.
[88] *Handelingen Tweede Kamer*, 1983–4, UCV 95, 14 May 1984, 1, 3, 6.
[89] Dr W. M. J. den Boer, interview.
[90] Professor N. van Breemen, interview.

visits were undoubtedly important for the creation of an image of damage in the inner circle of decision-makers and politicians.

Another prominent practice in the construction of a common image of damage and shared understanding of the problem was symposia. Especially significant in this respect was a major symposium on acid rain held at Den Bosch in November 1983. Interestingly, unlike the case at the British Royal Society meeting of September 1983, this meeting was attended not only by scientists, but also by politicians, representatives of employers organizations and trade unions, as well as spokespersons from industries concerned and agricultural organizations. They all expressed their concern.[91]

The symposium was the first of its kind.[92] Here Dr Pieter Winsemius, the minister for VROM, immediately precluded any antagonism between environmentalists and Government. The first lines of the opening address of Minister Winsemius were quoted at the beginning of Section 5.4. They left no room for doubt: the Government was facing the issue and argued that 'the problem of acidification [was] to be taken very seriously indeed.'[93] No NGO could accuse the Government of complacency any longer. On the other hand, it was a positional statement that challenged potentially affected groups. If one wanted to caution the Government not to make premature heroic efforts, this symposium would be the place. All interest groups had their representatives in the audience and the scientific cynics were present too. Yet despite the fact that the whole acid-rain problem was apparently so vague that it could only be formulated in terms of a 'flood of reports' on damaging effects and not in terms of directly observable degradation, none of them argued against the Ministerial position during the symposium. It illustrates that the actors knew only too well how politics is to be conducted in the particular culture of Dutch politics (see below).

[91] The symposium was organized by the newly founded *Vereniging Lucht* (Air Association). It may serve as an illustration of the inclusiveness of the Dutch debate on acid rain. In the preparatory committee we find both progressive politicians and representatives from environmental NGOs as well as representatives from the employers organization VNO/NCW and all relevant quangos.

[92] Other important but less public gatherings were the symposia at Twickel Castle in March 1983, April 1984, and March 1985. Here actors from various ministries, implicated industries, NGOs, and universities and research institutes came together to discuss the phenomenon of acid rain and the way in which it should be regulated. It was a case where VROM and NGOs co-operated closely.

[93] Winsemius 1984: 17–18.

Sanding Madonna: The Awareness Campaigns

A third important practice in the creation of an image of damage was the public awareness campaigns. Whereas in Britain the Government was reluctant to act, the Netherlands' policy-makers were trying to create the social support for abatement measures that would no longer be limited to industry but would also directly affect the public itself. Here VROM actively took on the task of reconstructing the image of damage. It staged a major Stop Acid Rain campaign in March 1985 and a second campaign Acid Rain—The Measures in April 1986.[94] The campaigns facilitated the growth of the general awareness of the complex origin and fundamental consequences of acid rain.[95] The campaign was a deliberate play on the apocalyptic sensibility of the Dutch. Its main instrument was the pictorial representation following the sequence 'Yesterday–Today–Tomorrow?' As Winsemius put it:

Most effective, I think, was our image of Mary. A little sculpture of a Madonna with little Jesus on the arm. We sanded it to imitate the effect of acid rain. That drove the message home. . . . We are ruining our society, all the things we value . . . and we do have a responsibility.[96]

VROM used pathos to reinforce the Christian image of 'being a guest in your own house.'[97] It was subsequently to produce the social basis for action. To enhance the public awareness of damage was not simply a matter of good information. As Winsemius

[94] The VROM campaign used leaflets and guides, billboards, television spots, and advertisements in the major Dutch dailies. Following the pattern 'Yesterday–Today–Tomorrow?' it showed forest damage (dying trees), damage to historic buildings (a crumbling sculpture of a Madonna with child), and animals and plants becoming extinct: see Anon. 1986*b*.

[95] To give an idea of the scope of these campaigns, during the six weeks of the campaign the public ordered a total of 700,000 full-colour guides explaining acid rain. The first print of 100,000 copies was sold out on the day the campaign was launched: see Anon. 1986*b*.

[96] Dr P. Winsemius, interview. This remark might suggest that the damage done to the cultural heritage was politically more important than the effects of acid rain on forests. This was most certainly not the case. Tree health was the issue on which arguments were tested, and which decided on the allocation of credibility: did the government take the appropriate precautionary steps, or was more action needed. Drawing on the format 'Yesterday–Today–Tomorrow?', environmental NGOs were quick to point out that for forests, moors, and nature in general 'there would not be a day after tomorrow', see De la Court *et al.* 1987: 4.

[97] This later became the title of Winsemius's acclaimed study on environmental management: see Winsemius 1986.

argued 'it was not so much the damage as the fact that you had a responsibility. Damage was the symptom of that responsibility.'[98]

For VROM the question was how to put acid rain at the top of the public agenda and to generate political support for the forthcoming policy package. Their strategic manipulation of the social perception of damage was firmly grounded in the concept of policy life cycles that Winsemius had made one of the organizing principles of his department. Basically he argued that successful crisis management required the distinction of certain stages, the first being recognition.[99]

We argued, if the policy life cycle is true, it must be possible to turn it round. It must be possible to give publicity to acid rain and make it into a huge political priority . . . We had long discussions on this matter at VROM. Should we really do this? It can break your neck. We then argued: 'Is the problem important?' Yes, they said. So we risked our neck . . . and started the big acid rain campaign.[100]

The motive behind the VROM campaign was Winsemius's wish to restructure the regulatory regime in pollution politics and to introduce the ideas and concepts of ecological modernization. Winsemius strongly disapproved of the traditional approach in Dutch environmental politics. For him pollution should not be seen as a political struggle between polluters and polluted but in good eco-modernist fashion: as a matter of innovation versus inefficiency. Whatever the purpose of the campaign appeared to others, its symbolism (the juxtaposition of pictures of fuming stacks of industry, of a car exhaust pipe, and of the rear end of a cow under the heading 'Every "exhaust" causes acid rain') certainly marked a shift away from the contradistinction polluter–polluted. It was a move towards the image of environmental decline as a common problem for which everyone had some responsibility. In this sense the new image of damage brought about a reconstruction of the air-pollution problem that did not allow accusations against others, but questioned the role of every citizen (as consumer, as motorist, or even as—responsible—parent).[101]

---

[98] Dr P. Winsemius, interview.
[99] See also Sect. 5.6 and Winsemius 1986: 15 ff.
[100] Dr P. Winsemius, interview.
[101] Interestingly, market research initiated by VROM indicated that the campaigns were only partly successful in changing the established perception of air pollution as an industry-related problem. Both traffic and agriculture were still seen as much less significant. See Anon. 1986b: 156.

The success of the campaign should be understood in its context. Obviously the government did not go against the grain of social feelings at that time. Neither was the government the only party responsible for the reconstruction of the air-pollution problem and the construction of acid rain. Radical NGOs had prepared the ground with media interventions, books, and excursions.[102] Yet whereas the NGO WISE managed to get 10,000 people into the woods in its most successful *Bosalarm* (Forest Alert) campaign, the departmental campaign reached almost every Dutch citizen.

This analysis of the creation of an image of damage leads to a rather unexpected conclusion which potentially has considerable consequences. It suggests that the construction of the acid-rain problem made it appear to be a much more unmanageable problem than it actually was. The repercussions of the high level of generality at which the tree health surveys were presented obscured the direct relationship between the ammonia emissions from intensive agriculture and the dying of forests. On top of that, it was the superbly simple run of the SPSS programme that illuminated this relationship. The publication of the SPSS results in 1984 could have been the basis for a focused effort 'to get the crocodiles out of the water' which was what Winsemius intended. However, here the interests of the Department of Agriculture and Fisheries clearly reinforced a very general representation of damage. Whether consciously or not, it was precisely this abstract construction of the problem that made it appear to be unmanageable. This was only reinforced by the general awareness campaigns of VROM.

Related to this finding is the observation that the whole debate took place under the shared assumption that dying trees should be seen as indicators of broader social developments. In this respect there was no attempt to defy the implicit social critique in the discourse on acid rain. For instance by arguing that if the trees did not grow one should change the crop. Or that it was a strange assumption that (imported) trees should be able to grow on the poor soils where they now happened to be. The latter position, that forestry experts—quasi-ironically—liked to raise in the interviews, was simply absent in the political debates. This was, as we will see, completely in contradiction to the various other ways in which policy-makers and politicians thought it acceptable that

[102] e.g. Van Ooyen and De la Court 1984.

nature should be manipulated to curb the damaging effects of acid rain. Yet trees were not crops but social indicators, and if they are meant to show the moral temperature of a society it is hardly legitimate to change the position of the thermometer to influence the outcome.

5.5. DUTCH SCIENCE AND POLITICS: CHANGING BOUNDARIES

> The details, they became black boxes for me, I did not need them. I only had my input and output. Only when it was necessary to provide the levers for policy-making did we get into these details.[103]
>
> Dr T. Schneider, programme director, AVO

Almost inevitably science plays a key role in defining regulatory strategies in environmental matters. So too in the case of the Dutch regulation of acid rain. Yet whereas in Britain policy-related science focused on the question whether or not acid rain was a serious issue, Dutch scientific input took that as given and immediately tried to find the critical limits of what nature could endure instead. Yet one of the interesting facts of the Dutch debate was that as the acid rain game unfolded, the positioning of science changed. At first scientists worked happily within the confines of the positive-sum game format underlying the moderate Dutch discourse-coalition on ecological modernization. As within a couple of years the limits to this positive management approach became more obvious, science took on a new role. With the RIVM publication *Zorgen voor morgen* in December 1988, science introduced a reflexive moment to (air-)pollution discourse, exposing the limits to the Dutch discourse of ecological modernization. *Zorgen voor morgen* suggested the need for structural change and paved the way for a more radical *Nationaal milieubeleidsplan*, which was to be the political answer to the perceived failure to put the eco-modernist ideas to work. In all this the experience of scientists in the acid-rain controversy played an essential part.

---

[103] Dr T. Schneider, interview.

The AVO Project or the Search for Critical Loads

In the acid-rain case science did not gain access to the policy domain until 1982. Dutch environmental policy-makers knew about the Scandinavian reports on damage but were not really aware of the work of the few Dutch researchers who had already published on the effects of acidification in the Netherlands. In the early 1980s, more scientists became concerned. The acid-rain threat became more manifest as researchers found evidence that ammonia, until then only classified as an obvious nuisance in agricultural regions, was related to acidification.[104] While the return of Minister Lambers-Hacquebard from Stockholm denoted the political move from an SO_2 story-line to the acid-rain story-line, it was the motion De Boois of February 1983 that corrected the bias in science. Significantly, the motion asked for research to be conducted on soil acidification, not acid rain. It thus modified the politically induced bias in acid-rain research and brought ammonia into the discussion.[105]

The motion facilitated the work of scientists who wanted to do research on air pollution in a more inclusive way. Many of them had only been waiting for the go-ahead. Within six months the motion resulted in a substantial literature survey and within a year the so-called *Evaluation Report* or *IWACO report* was published. The scientific report suggested that the natural damage called for more stringent measures and illuminated the deficiencies of the traditional regulatory practice of defining national emission ceilings.

Once this report had been accepted, the policy-maker's concern was to come up with the appropriate response. What would be considered appropriate was rather hard to assess for the policy-makers in the Air Division. After all, VROM was reconsidering its entire policy regime at that time. Nevertheless, given the neo-liberal commitments of the Lubbers Cabinet, policy-makers realized

---

[104] Most prominently Dr W. Asman and Professor N. van Breemen (see van Breemen *et al.* 1982).

[105] Dr De Boois explained her decision to call for research into soil acidification by referring to her own Ph.D. research on the development of microsis and fungi in the forest. 'I had seen how the condition of the forest was getting worse. So I knew, whatever limit you set, it will be too high. . . . The pollution of the soil is irreversible. If you stop polluting, the air will be clean. Water can also become clean again. But soil is different. You really are dependent on that topsoil.' (Dr R. de Boois, interview).

that credible control of pollution had, more than ever, to avoid unnecessary societal upsets and had to seek to maximize the effectiveness and efficiency of its regulatory effort. Such were the boundaries within which the discourse of ecological modernization had to be developed. Apocalypse might dominate the public debate but policy-makers would ultimately have to legitimate their proposals in the discourse of efficiency and science. Therefore they needed precise knowledge of the relative contributions of various emissions. Moreover, if the Air Division wanted to maximize its effectiveness and avoid unnecessary social upset, it needed to know at what level of emissions damage would start to occur.

This was where the practice of 'critical loads' came in. The concept was strongly promoted by environmentalists but in fact the philosophy behind the concept was not new. In public health policy the approach had always been to try to find the critical level (referring to certain concentrations) at which specific negative effects to humans would no longer occur. Policy-makers now transposed the idea to environmental policy. What was new was the application to nature ('load' referring to the exposure or deposition), something which had been pioneered by the Swedish government in the context of the acid-rain problem. A first attempt to define a critical load regarding acidification can be found in the IMP-air 1984–1988.[106] This first definition of a critical load was more a guestimate than an estimate: 'One person in the Department assembled some literature from Canada and Sweden and then we simply estimated the difference in soil conditions in Sweden and here and came up with a number'.[107]

To determine a critical load one needed, once again, an idea of what was natural, but in Dutch policy-making the awareness that the country was one big human construct was prevalent. Nevertheless, policy-makers had to choose what society should seek to protect: do you want lichens to live everywhere, do you aim at perfect fishery conditions in lakes, or do you aim to guarantee the provision of enough recreational forests of whatever sort grows best? Such points of reference then subsequently often become the social construct of the natural. We decided to focus on something

---

[106] *Handelingen Tweede Kamer*, 1983–4, 18 100: 69–70.

[107] M. Bovenkerk, VROM Air Division, interview. Another respondent remembered: 'That first calculation was complete nonsense but coincidentally not far from the truth as we found later on.' (Professor N. van Breemen, interview).

that the public is attached to, something that was socially important. That was the forest. We argued it should be possible to maintain healthy forests in the Netherlands at the places where they are right now [i.e. mainly on poor sandy soils].[108] Once nature was socially constructed science came back in. Since critical loads had been agreed as the point of reference for environmental policy, a more robust scientific basis was required. After all, the critical load would determine the scope of the regulatory task and the required emission reductions. It is in this context that we have to understand the role and function of the *Additioneel programma verzuringsonderzoek* (Dutch Priority Programme on Acidification, AVO). AVO was the Dutch counterpart of the British SWAP research, and, in many respects, illuminated the biases of Dutch environmental discourse just as SWAP reflected the British predispositions.

AVO had its roots in the big acid-rain symposium in Den Bosch in November 1983. The organization of the Den Bosch symposium already indicates that Dutch acid-rain science was not so much concerned with the question 'Is it true?' as with the question 'What should be done?' At the time even vested interests like Shell (refineries) or the electricity utility SEP (coal-fired power-stations) had accepted the seriousness of the problem.[109] The broadly shared consensus at Den Bosch predetermined their attitude to the government's regulatory approach. In December 1983 the head of the environmental division of Shell, Mr van Lookeren Campagne, took the initiative to stage a meeting at the Holiday Inn hotel at Leiden. He invited the directors of the main research institutes as well as the leaders from the implicated industries. Eventually, the Holiday Inn meeting led to the initiation of the AVO research.

Officially, the AVO research[110] was first announced in the *Onderzoeksnotitie* (Research Memorandum) of June 1984[111] and

---

[108] M. Bovenkerk, interview.

[109] This perception was most certainly reinforced by the visit of representatives from these industries to the German forests in the autumn of 1983. It had made them aware that the issue was too serious to be denied. They were shown around by Dr Mauk den Boer (Staatsbosbeheer) and the director of the forestry research institute *De Dorschkamp*, Dr van den Bos.

[110] 'Additional' in AVO refers to the fact that at the time research was already under way with a value of about 40 m. Guilders, most of which went into experimental set-ups for SO_2 and NO_x abatement. These pilot projects were part of the regulatory effort initiated in response to the SO_2 story-line.

[111] *Handelingen Tweede Kamer*, 1983–4, Onderzoeksnotitie, 18 224, no. 18.

was inaugurated in January 1985.[112] In AVO the ministries of VROM, Economic Affairs, and Agriculture and Fisheries co-operated with the refineries and with SEP. Each party contributed one million Guilders annually for a three-year period. The actual research management was given to a *Projectgroep* made up of scientific experts[113] led by Dr T. Schneider from RIVM. The AVO project should come up with scientific answers regarding the effect-iveness of possible abatement measures.

It is not entirely obvious that AVO was the intended outcome of the Holiday Inn meeting. As one of the key participants recalled:

At that time there were some forces within these firms that wanted to examine whether it was possible to bury the whole issue. Over these Holiday Inn meetings there was a veil over what is going on, what is the underlying intention of the initiators. But quite quickly we found one another in a common effort on the issue. It might well be that they realised that the Government was determined to keep the momentum going and that the different departments, Economic Affairs, Agriculture and VROM were in close contact on this matter.[114]

It seems quite likely that the parties tried to assess the climate and secure an influential position in the production of knowledge that would set the boundaries for new regulatory commitments. As the president-director of Shell Nederland argued 'Our real motivation was based on the recognition that something started to move. We better make sure we are part of this process rather than letting it run its course.'[115] Obviously, the participation of industry in science partly reflected a concern that action would be taken too quickly. Financial participation in AVO research gave the indus-tries the chance to monitor the research closely and to control the interpretation of findings. Yet it is interesting to note that both

---

[112] Its rather complex organizational structure reflected a year of political bar-gaining. In the end the AVO project was governed by a steering group made up of representatives from the ministries and industry, while the research was led by a project group at the RIVM.

[113] Originally there was also a third council, the so called *Voorportaal*, a pre-paratory committee. It was the outcome of the negotiations (each ministry would be chairing one committee), but quickly became redundant.

[114] Dr B. C. J. Zoeteman, interview. Zoeteman, who is now Deputy Director-General at VROM, was involved in the Holiday Inn meetings as research director at RIVM.

[115] P. van Duursen, interview.

Shell and SEP were willing to participate in AVO without being able to phrase the research questions in advance. Unlike the CEGB in the SWAP project, the Dutch industries participated in a round-table discussion together with departments and research institutes to come to a description of the relevant questions. It was agreed that AVO should answer the following questions: Which substances are responsible for the damage caused to vegetation? How (by what means and in what way) is this damage inflicted? How effective are abatement measures?[116]

AVO illustrates, once again, the integrative or centripetal characteristics of Dutch environmental policy-making. Acidification was quickly recognized as an issue that could well be the basis for an antagonistic public debate, and promptly became an issue for private negotiations among leaders of science and society. What is more, AVO became a practice in which Government, industry, and science co-operated to find solutions that would not unnecessarily put the existing industrial practices, whether it was in electricity production, oil refinery, or agriculture, in jeopardy. In other words, it structured the positive discursive regime of ecological modernization that was being laid out in the Winsemius era.[117]

AVO published its first results in the *Tussentijdse evaluatie* (Interim Evaluation) of August 1987.[118] The Project Group argued that a continuation of present deposition would result in further decline of forest vitality and further transformation of heathland into grassland. It argued that a reduction of deposition of up to 90 per cent was required to restore the natural state of forests and heaths (but also to comply with the EC norms on nitrates in drinking water). It presented its first calculation of the critical load expressed in so-called 'acid equivalents'. Acid equivalents were obviously an aggregate measure. They served to provide a common denominator that allowed the comparison of the relative contribution of different pollutants (such as SO_2, NO_x, and NH_3). At that time the total emissions were calculated to be about 5,800 acid equivalents and the official policy target was an emission of 3,000 acid equivalents in the year 2000. The AVO research suggested, however, that a deposition of 1,400 eq. of 'effective acid'

---

[116] Stuurgroep Verzuringsonderzoek 1985.
[117] We will come back to this 'positive management approach' below.
[118] Schneider and Bresser 1987.

would still result in serious damage.[119] Critical load should be 1,400 equivalents of potential acid for heathlands and coniferous forest and 2,400 eq. of potential acid for non-coniferous forests. The ozone levels should not exceed the natural levels. Obviously, these findings called for profound changes in government policy.

In November 1988 AVO published the final report of the first three-year phase of the project.[120] Calculations had shown that the critical load level was in fact even lower than previously thought. The critical load was now thought to be about 400 to 700 acid equivalents annually per hectare. This was, in fact, roughly similar to the natural deposition levels. Hence the rather ironic outcome of a project to find the critical limits up to which pollution could be tolerated was that, given the goals of Government, in actual fact no pollution would be legitimate at all. Quite unintentionally, AVO led to the periodical reinforcement of a legitimation crisis: the Government was not doing enough to safeguard the environment.

How did all this come about? The official task of the AVO project was to facilitate the closure of the problem of acid rain, i.e. to find practical solutions. Yet this task was of course beset with problems. Only a simplistic view of the science–policy interface would maintain that this was a matter of first getting the facts right and then reporting on possible trajectories of action. In actual fact scientists, science managers, and policy-makers first had to carve out the respective domains of science and policy and define the appropriate relationships between them.[121] What is more, as the above already suggests, this process had its own dynamic and could, in fact, only be partially controlled by the various participants.

Bring in the Model-Makers!

In many respects AVO was a unique attempt to reconstruct the role of pollution science in the context of a discourse-coalition

---

[119] This was a complex theory. Until then the influence of nitrogen had been underestimated. Because of new scientific findings the 1,600 equivalents of nitrogen emissions that had until then been seen as 'benign' had to be included in the calculation as well. This is where the distinction between 'effective' and 'potential' acid comes from. In its initial 'guestimate' the government had suggested a level of 1,800 acid equivalents as a very preliminary indicator of a level below which no acidification would occur.

[120] Schneider and Bresser 1988.

[121] On this subject, see Jasanoff 1990; Wynne 1992a.

organized around ideas of ecological modernization. AVO followed the earlier attempt to represent the problem of acidification in the IWACO study. Yet while the IWACO report was based on five separate papers that were drawn up very much according to the disciplinary division in science,[122] AVO sought to come to a truly interdisciplinary understanding of acidification. First implicitly, later explicitly, it conceptualized nature in terms of ecosystems and made individual scientists, whether working on aquatic ecosystems, tree physiology, or atmospheric chemistry, work together in order to generate knowledge at the level of the ecosystem. Yet an integrated research approach called for more than ontological changes alone. AVO had to find a practical model of communication that would allow for the meaningful relation of knowledge produced in disciplines as varied as, for instance, biology, chemistry, physics, forestry, and soil science. How do you bring about this interdisciplinary form of communication? One of the research directors involved argued 'In actual practice there are two ways to arrive on a really integrated approach: you choose a common object (i.e. send everyone into the same forest) or you have everybody dancing around the same theoretical model. We did both.'[123] Researchers agreed that AVO management was remarkably effective in organizing a productive research community out of the 30 research institutes that took part in the project. In order to achieve this the research management choose not to aim at bringing together the most excellent researchers but sought to involve people who wanted to communicate and co-operate.[124] Hence it recognized the peculiarities of pollution science and rather than completely imitate the model of pure science it found its own ways of creating a regulatory science setting in the sense of Jasanoff.[125] Within three years the integrative strategy of AVO produced a research community on the theme of acidification based on mutual trust.[126] 'All pieces of evidence contributed to the common answer. One even used one another's conclusions for one's own projects.'[127] It even caused researchers who had their own funding and were working outside the direct confines of the AVO

---

[122] The five studies were: atmospheric processes and deposition, soil, soil biology, vegetation, and surface water and hydrobiology. On the issue of damage to historic buildings, etc. a separate report was drawn up. See Feenstra 1982.
[123] J. C. A. M. Bervaes, interview.   [124] Dr T. Schneider, interview.
[125] See Sect. 4.6.   [126] Dr R. M. van Aalst, interview.
[127] A. H. M. Bresser, interview.

project to restructure their work to be able to take part in the AVO debate. This interdisciplinary work resulted in what one could call, using a term from the discourse of ecology, scientific synergism: once mutual trust emerged, all kind of new insights emerged from the interaction among researchers that each perceived reality in their own particular way.

I think every consistent description of the problem shows the need for coordination. That means that you cannot act simply as a plant physiologist or an air quality specialist. It simply becomes clear that the surplus value of putting your data next to the data of somebody else is so big that it is in your own interest to do so.[128]

One of the differences with the British situation was the fact that the individual projects were consciously related to one another (also because of the decision to focus the research work on a few field-study areas). Consequently, the Dutch research held together well.[129]

A second factor that boosted the effectiveness of the AVO project was the development of the so-called DAS-model (Dutch Acidification Systems). The DAS-model was basically an advanced mathematical version of the familiar conceptual models of pollution (anthropogenic emission, transport, deposition, effects).[130] It tried to quantify all input, throughput, and output and tried to model the chemical interaction during this process. Hence the DAS model is not to be understood as a causal, analytical model, it was more a descriptive input–output model that owed much to the science of cybernetics.[131] Initially, the programme directors were not sure DAS would be a success.[132] Yet later the modelling got to play a

[128] Dr R. M. van Aalst, interview.

[129] As far as the scientific standard was concerned, the international review team argued that it was impressed with the overall standard of AVO research: see Schneider and Heij 1991. One of the possible deficiencies of the AVO project might be the strong emphasis on work in the field. AVO paid far less attention to actual lab research. Modelling was mostly based on field measurements, but these were not often reproduced under controlled lab circumstances (Professor N. van Breemen, interview).

[130] Conceptually DAS can be compared with the RAINS model of IIASA (Austria). The soil module of DAS was actually taken over in the RAINS model.

[131] In actual fact RIVN was in close contact with foreign specialists in cybernetics to solve conceptual problems (A. H. M. Bresser, interview).

[132] Dr T. Schneider, the programme director, argued: 'If I am honest, I have had serious doubts whether it was essential.' (interview).

central role in the AVO project.[133] This was because DAS provided the interdisciplinary research with a kind of systemic meta-language that helped to overcome disciplinary biases. The DAS model forced every researcher to see whether the system description matched his or her own findings or hypotheses. It thus also functioned as a discursive practice that stimulated scientists to engage in a wider debate on the level of the description of acidification as a systems phenomenon.

As the reluctant reception of DAS by the AVO research directorship already indicated, the idea of using large-scale modelling for the regulation of an environmental problem was not self-evident, especially not for scientists. Yet in actual fact DAS is a remarkable illustration of the influence of engineering discourse on the Dutch science of environmental regulation. As I found out, DAS drew heavily on the established expertise of Dutch projects in water management. Especially significant in this respect were the PAWN model (Policy Analysis of Water Management in the Netherlands) of *Rijkswaterstaat*, the study into drinking-water supply in the Zuid-Holland province, and the Grevelingen and Oosterschelde models that had been developed in the context of the *Deltaplan*. These were examples of projects that aimed to come up with both an understanding of nature and prescriptive management strategies. Some of the key individuals in the management of the AVO project (e.g. F. Langeweg and A. H. M. Bresser) had previously worked on these comprehensive multi-disciplinary projects.[134] 'You recognised that a certain approach worked so you apply it again, in another context. The tools that we used there were no longer useful but the philosophy was most certainly drawn upon in this other project.'[135]

---

[133] The theoretical model was recognized as a 'unifying force' by the review group (see Schneider and Heij 1991). Participating researchers also emphasized the importance of this mathematical model: 'It really was what held much of the work together. It forced the programme directors and researchers to fill in the gaps, to work on understanding the missing links in the chains.' (Professor N. van Breemen, interview).

[134] The researchers involved in the system-analytical approach of the AVO research were mostly civil engineers from the Technical University of Delft. The disciplinary background of the model builders in Dutch environmental research in fact varied from civil engineering to acoustics. Together with ecologists (another brand of system thinkers) they provided the strong systems analytical input into the understanding of acidification.

[135] A. H. M. Bresser, interview.

It was, in other words, a paradigm example of the 'garbage-can' model of a solution looking for problems.[136] Indeed, nowadays the same engineers use their expertise in the context of the policy-oriented research on the greenhouse effect. DAS illustrates the practical way in which the Dutch preoccupation with water management filters through to adjacent domains of regulation: it is a well-institutionalized, well-legitimated pool of expertise. DAS also signifies the influence of engineers on the actual research design. This is most relevant since engineers, and particularly the civil engineers who had their backgrounds in the large *Deltaplan* project, have a rather pragmatic approach to science and policy. For them knowledge of nature's economy is not a goal in itself: knowledge is there to help manipulate nature to the benefit of mankind.

Obviously, the aggregation of knowledge from various different disciplines came at a cost. The AVO project and the central DAS-model rested on a rather arbitrary reduction of the actual complexity of nature. Firstly, AVO started from observations on a few field sites. The information from the fieldwork was then used to design a computer model. The next move was to extrapolate from that model to be able to account for regional differences. Yet, as many researchers indicated, the two sites that were chosen as research areas were compromises that later came out to be less fortunate choices. Secondly, the AVO research was restricted to the policy problem of acidification and attempted to compare all kind of different factors in its explanation of this single abstract process. Yet, as forestry experts constantly emphasized, tree health or forest decline is in fact a more complicated issue. Insects, fungi, droughts, and frost all have their impact, yet it is extremely hard to account for these incidental complications. To the annoyance of forestry experts these factors had to be forced into the logic of acid equivalents, which was, after all, the denominator of the whole project. Thirdly, AVO was directed to measure growth as an indicator of health, but at various places nitrogen was found to stimulate growth. Nevertheless, AVO clearly was successful in creating an interdisciplinary form of scientific communication.

A third important element of AVO (next to the creation of a research community and the DAS model) was the intermediary

---

[136] For this garbage-can model of the political process, see Cohen *et al.* 1972; Kingdon 1984.

role of the programme managers. Obviously, scientists and policy-makers tend to see the exercise in an entirely different way. The policy-makers believe in the metaphor of filling in the gaps of knowledge while if scientists really start to look into things they are likely to argue that they knew less than they originally thought.

A few years ago one could argue that something was perhaps not well understood yet but that the state of the environment was worrisome. Now we have done much more work and have found that it is impossible to come up with straight cause–effect relationships between deposition and damage.[137]

What is needed in such exercises is an intermediary, a knowledge broker, who is able to relate these different (composite) discourses and come to some form of closure. In the Dutch case this role was played by the programme management of the AVO project. According to the director of AVO research, Dr T. Schneider, the main question for the policy-maker was 'which are the important pollutants and what are the effects? In other words, what should we monitor, once we start to turn the taps to see whether our actions have any effect? Thirdly, how effective is our action?'[138] At the other end stood the researchers. They had no interest in the compilation of policy relevant data.

Critical load does not have any meaning for me. I do not consider it a scientific question. But I understand that it is relevant for policy-makers. . . . We say it is too much, they want to know how far they should go because we cannot go back to the middle ages. But this is not a scientific question.[139]

The research management thus had to force problem closure. Here the modelling of nature through DAS played a key role. For the AVO management the detailed mechanisms that scientists were interested in were less relevant than the observed empirical regularities that the scientists used as the basis for their own experimental work. According to Schneider 'The details, they became black boxes for me, I did not need them. I only had my input and output. Only when it was necessary to provide the levers for policy-making did we get into these details.'[140] In this way the modelling

---

[137] Dr R. M. van Aalst, interview.   [138] Dr T. Schneider, interview.
[139] Professor N. van Breemen, interview.   [140] Dr T. Scheider, interview.

was a very instrumental practice for an effective form of eco-modernist environmental management.

Controlling the Mediation of Knowledge

In the end AVO results failed to set a change of policy in motion. This can be partly explained in terms of the institutional set-up for the project. The programme managers of the AVO project were not in total control of the translation of knowledge into policy-relevant information. They were part of the Project Group of AVO and acted as knowledge brokers, but were not policy-brokers.[141] In actual fact there is ample evidence for the assertion that the Steering Group, in which ministries and industries were represented, at various points acted as gatekeeper and thus inhibited a more effective application of the knowledge acquired in the context of AVO.

For instance, in the Interim Evaluation of August 1987, the Steering Group in a way tried to play down the immediate political significance of the figures. It cautioned that there was an uncertainty margin of 40 per cent and furthermore argued that

in order to be able to come to well-founded conclusions with regard to the adjustment of deposition targets, a better idea is required regarding the degree to which uncertainties in measurements, calculations, dose–effect relationships and in research findings and extrapolations, affect the accuracy and robustness of the final results. Furthermore a judgement on the criteria used, requires a good description of damage effects, the possibility for repair and the societal meaning of the damage caused.

The Steering Group added that

Stress factors such as frost, drought, illnesses, and plagues, as well as factors such as location, maintenance, choice of type of tree, and ground water drainage can also greatly influence the reported damage. Those factors that are connected to human action ought to be included in the abatement strategy.[142]

These arguments are reminiscent of the traditional-pragmatist story-line in the British controversy. Whatever were the Steering Group motives, it obviously contradicts the findings of, for instance, Dr

---

[141] On the concept of policy-brokers, see Sabatier 1987.
[142] Schneider and Bresser 1987: 5.

den Boer, who had by then already established direct correlations between certain emissions and forest damage in 1984. Furthermore, it seems that a further delay of remedial action was hardly defensible given the discrepancy between the actual emission of 5,800 equivalents and the estimated critical load of 1,400.

Another element is the artificial creation of a distinction between science and advice. Originally the Project Group had included recommendations on what measures it considered most effective. This was a consequence of the dynamics of both the DAS model and the engineering culture. The whole modelling effort started with an 'input' of certain emissions and it was rather obvious that the pragmatic engineering approach would result in trying to assess what regulatory efforts would come out to be most effective according to the model. In the event, the Steering Group refused the draft and ordained a more analytical perspective. In other words, the Steering Group pushed back the domain of science in the direction of pure research and thus prevented the full application of the available instruments on the closure of the problem.

This failure to come to problem closure on the basis of available scientific knowledge was not an incidental but a structural feature of Dutch pollution science. We already encountered it in the analysis of the tree vitality surveys but in AVO it clearly featured prominently as well. AVO had given persuasive evidence to support a tight regulatory regime; actual deposition was shown to be about 6,000 acid equivalents while the critical load was perhaps about 1,500 or even 400 to 700. In other words, the actual deposition was always two, three, or more times higher than the deposition that would keep a forest or a heath terrain in good condition. One could argue that policy-makers therefore had ample arguments to come to further closure of the problem.

The Maintainance of Complexity

What happened instead was that the intricate discursive instruments that were basically heuristic devices, came to play a role for themselves in political discourse. Concepts like 'acid equivalents', 'critical load', and the categorization of tree vitality became an independent level in the political debate on acid rain that shielded the coupling of problem perception and problem closure. Here the numerical representation, that had been meant to set the parameters of the regulatory debate, started to obscure the character of the problem itself:

At some point this quest for numbers and those numbers seem to have become a much more important basis for policy than, say an intellectual persuasion, a qualitative approach . . . Numbers became sacred. Nobody spoke any longer about the 'firmness' of a figure, the degrees of uncertainty under which these figures were arrived at.[143]

The seemingly uncontested credibility of science was also noticed by one of the key policy-makers in the Air Division of VROM: 'I have always been surprised that target groups always took these figures [critical loads] for granted. They always argued about the relative contribution of their target group never about what is the basis of that critical load, what do you want to preserve in nature etc.'[144] In fact critical loads were not the basis of the political deliberations but came to be an element in argumentative carousel in which many other variables circulated. Actors would all agree that serious damage was unacceptable but were, at the same time, willing to take some interim goal (or target load) as the starting-point of their negotiations. This made the whole scientific effort to define as precisely as possible the uptake capacity of nature into a mere legitimating effort. Dutch critical loads were not the basis for the social definition of a sustainable solution to the acid-rain problem. As a consequence the Government was in the bizarre position of knowing the precise critical load levels while it argued, at the same time, that they could not be attained within two decades.

Only the most radical NGOs refused to enter into a negotiation that took the deliberate failure to meet generally agreed upon and publicly endorsed goals of Government policy (preventing the worst damage) for granted:

That whole discussion on acid equivalents was a non-debate. The fact was that we had a deposition of 8,000 equivalents whereas the natural deposition was about 400 to 800. That was what we argued. We thus said something about anthropogenic emissions, over our role as human beings in this world. As soon as you get close to 1,400 you start to manipulate in a technocratic way.[145]

---

[143] J. C. A. M. Bervaes, interview.
[144] M. Bovenkerk, interview.   [145] T. de la Court, interview.

The Dutch regulatory regime thus introduced eco-modernist notions but resisted the technocratic regulatory regime and stuck to the peculiar Dutch tradition of accommodation. It focused political discourse on the negotiation of trade-offs between parties involved but without taking the scientific assessment of the situation as a starting-point.

The failure to use the available knowledge to come to problem closure is also evident in the failure to break down the abstract representation of acid rain. One could argue that the finding that only a 90 per cent reduction could prevent further damage, makes the task of regulating acid rain look almost hopeless. So if AVO could have illuminated the precise points of application that Winsemius wanted to know, if the regulatory scientific project could have indicated how one could have taken the 'big crocodiles out of the water', and at what cost, it would have contributed to linking the level of problem perception with the level of solutions. Yet by sticking to the analytical information, AVO failed to break through the characteristic discursive split in a realm of problem perception that was overdetermined by apocalyptic constructs and a realm of problem closure that was overdetermined by economic considerations. A discursive split that seems to be the single most remarkable feature of Dutch environmental discourse.

However, the fact that the accommodative regulatory regime imposed certain limits on the role of AVO in the regulation of acid rain should not lead to a purely reductionist explanation or an instrumental perception of the AVO project. In many respects, the dynamics within AVO reinforced this failure to come to closure of the problem. Basically AVO was a technocratic project. Sometimes the instruments (especially the DAS-model) became more important than the purpose (what should we do about acidification). As one of the researchers involved argued 'These research projects with all those mixes of pollutants and what have you, that whole blurred affair and all that calculation and translation, equivalents, etc. Every time ammonia fell between the chairs.'[146] While at various moments researchers found conclusive evidence of the essential role of ammonia in forest dieback, research and modelling continued aiming for an ever more refined representation

---

[146] J. C. A. M. Bervaes, interview.

of reality. In this sense, AVO itself contributed to the failure to come to effective problem closure.

Zorgen voor morgen *or the Redefinition of the Role of Scientific Expertise*

An account of AVO alone cannot give an understanding of the workings of the Dutch science–politics interface in full. The subordination of science to the technocratic search for limits had been a thorn in the side of many scientists, including many of those who had been active in AVO. It should be remembered in this context that in the Netherlands science was not nearly as much a self-contained community as in Britain. Not only was much of the relevant scientific expertise located in large research institutes that were officially working for policy-makers (such as the Rijksinstituut voor Volksgezondheid en Milieuhygiëne (RIVM) or the Rijks-instituut voor Natuurbeheer (RIN)), many scientists also had very close relationships with the environmental movement.[147] Other key scientists had direct relationships with the press.[148] For instance, when Professor van Breemen published his famous article on the role of ammonia in the acidification process in *Nature* he also compiled a press release to accompany his article:

I just said this and this is going on. The next moment you are approached by *Natuur & Milieu*, and the press and then you are part of the circuit. . . . I think the Netherlands has a culture that if you find something you think is important for the public to know, you go to the press. That is a service to the public but it is also in your own interest. Simply because you enlarge your own profile and improve your chances for recognition and financing.[149]

This quotation is interesting for two reasons. First, it illustrates the relative absence of a numinous form of legitimacy and the

---

[147] Jan Roelofs, an aquatic biologist from the University of Nijmegen, who was seen as one of the key scientists in the research on acidification, combined his academic work with a public campaign against acid rain, often in co-operation with NGOs like Natuur & Milieu or Milieudefensie. Willem Asman, one of the international pioneers on the role of ammonia in the acidification process, was at the same time a representative for the NGO Stichtse Milieufederatie to the European NGO Conference on Acid Rain at Göteborg, Sweden in May 1981.

[148] See the account of tree health.

[149] Professor van Breemen, interview.

weight of civil legitimacy in Dutch regulatory science. After all, the publication of scientific results in the popular press might, in other cultural circumstances, be exactly the way to lose your status as scientist. Second, this perception of the social responsibility of scientists was widely felt. It should not come as a surprise, therefore, that many scientists were also ill at ease with the representation of the significance of their work through AVO. This frustration was a breeding-ground for a much more wide-ranging statement on the ecological crisis. This new role for science materialized in the publication of *Zorgen voor morgen* ('In Care of Tomorrow'). The Department for VROM had asked the RIVM to come up with an authoritative assessment of the future of the environment and the repercussions of the various policy alternatives. What would happen if we did not act? Which alternatives can be distinguished?

The statistics contained in *Zorgen voor morgen* made overtly apparent that the constant increase in production and consumption volumes completely nullified the effect of the measures that were taken or being considered to combat acid rain. A much tougher policy package was required to achieve the government goal of preventing the worst damage. In essence *Zorgen voor morgen* combined the demonstration of the failure of the policies of the past with a presentation of the estimated costs of additional abatement measures. The message could not be missed: the weak operationalization of ecological modernization was not delivering the goods.

Although the RIVM research was not originally commissioned as part of a VROM masterplan to prepare the socio-political ground for the *Nationaal milieubeleidsplan,* it came to fulfil the role of catalyst. The intensity of the social outcry that followed its publication in December 1988 was only parallelled by the response to *Limits to Growth* in 1972. In many senses it was the conscious construction of a crisis. The play with apocalyptic rhetoric was hard to miss. *Zorgen voor morgen* contained graphs that shot up or fell steeply.[150] As the author of the section on acidification argued in 1991

Let's face it, in the end the goal of RIVM is to provide support for the policies of VROM. If you used cautious wordings on environmental matters

[150] See Langeweg 1988: 6.

a few years ago, nothing happened. So with *Zorgen voor morgen* we used some, let's say, somewhat 'stronger' language.[151]

The data on acidification that were presented in *Zorgen voor morgen* came straight from the AVO research. Inevitably, some of the authors were researchers for AVO. Yet here they did not position themselves as scientists, but as policy advisors and changed their discourse accordingly: 'We wrote this as policy advisors. That calls for a different language and a different presentation. It was more conclusive, more determined.'[152] While AVO was a scientific report that did not speak out on the social repercussions of its findings, *Zorgen voor morgen* allowed researchers to extrapolate from their scientific findings. So here science took up the role of policy-broker and introduced its own ideas about problem closure in policy discourse. For this purpose *Zorgen voor morgen* broke down the aggregate representation of acidification of AVO arguing that 'It is important to focus attention on those areas with sensitive ecosystems' and concluding that it was precisely there that 'depositions are higher than the national average.'[153]

On the other hand *Zorgen voor morgen* nevertheless carefully positioned itself within the accepted eco-modernist discourse of the government at that time. Hence it referred to the OECD report *Facing the Futures* and the UN report *Our Common Future* and explicitly subscribed to the belief in the regulatory capacity of technological innovation and social adjustment. In its radicalism it thus aimed to stretch the discursive confines of ecological modernization by pushing from within rather than pulling from outside. The extrapolated proposals for change were less radical than one would perhaps have expected given the convincing (scientific) image of the coming of an unprecedented crisis. In its discussion on what should be done, energy conservation was left undiscussed, as well as the option of reducing the volume of traffic while the rather structural idea of reducing cattle stock was mentioned amongst various technical abatement measures.[154] The report did not substantiate this choice for more or less pragmatic policy alternatives. Yet this was the result of strategic considerations. Former minister Nijpels argued:

---

[151] A. H. M. Bresser, interview.   [152] Ibid.
[153] Langeweg 1988: 110.   [154] Ibid. 429–30.

Zorgen voor morgen was very consciously released. It was meant to prevent the discussion on the NMP from foundering on all kinds of particularistic scientific disputes or on arguments from other departments that we had the facts wrong. It had to be an objective state of the environment which would allow for a debate on the measures that were to be taken but would fix the facts. You can then have a debate in which you say well we need so much reduction and then we can take the political decision whether you want to spend that money.[155]

Yet here the goal of accommodation directly determined the nature of the proposed policy package. 'If a report does not match reality or suggests such extreme alternatives it will be killed. Then it would have lost its function. In fact, it hit everybody like a bomb and everybody rushed, sometimes unwillingly, to subscribe to its conclusion.'[156]

This acceptance of the dominant frame also comes out in the concepts the report used to describe the environmental crisis. Although the report was explicitly not meant to be a scientific report but policy advice, it was still totally dominated by natural science discourse. Society was not analysed as a social system which functions in an environmentally malignant way. On the contrary: 'Society can be interpreted as a system of transformation of matter and energy, that are drawn from the physical environment.'[157] This was not so much a conscious choice of certain actors as the result of the prevailing cognitive bias that governed a research institute like RIVM. Although *Zorgen voor morgen* was the product of an interdisciplinary approach to the environmental dilemma, it thus did not include an analysis of the cultural, political, and economic practices through which pollution was produced, yet these were precisely the categories that would show why certain critical loads had been unattainable in the past.[158] This discursive

---

[155] E. H. T. M. Nijpels, interview.
[156] Ibid.   [157] Langeweg 1988: 12.
[158] One of the most influential participants in the drawing up of *Zorgen voor Morgen*, Dr B. C. J. Zoeteman, then research director at the RIVM, most certainly had his own ideas about the social dimension of the environmental crisis. Elsewhere he wrote about the social production of pollution in the following terms: 'Is man active in destroying the global physiological mechanisms of control? ... With the coming of modern physics and the industrial revolution in the 17th and 18th Century everything changes. The agricultural structure is broken, cities grow like cancers in a sick body and all flows of matter speed up and increase . . .' (Zoeteman, 1988: 71). Zoeteman was heavily influenced by the Gaia hypothesis. This also came out in the ordering principle of the report *Zorgen voor Morgen* following the

bias towards a natural science representation of environmental reality was subsequently reproduced in the NMP of May 1989.[159]

In conclusion one can say that AVO was a practice that upheld the modest positive regulatory regime that Winsemius had introduced as ecological modernization in 1984. The history of AVO on the other hand shows that the quick move of potential opponents to a radical regime of environmental regulation, such as the refineries and the electricity utilities united in the SEP, to take part in the new coalition, influenced the actual content of the terms. Through their activities in AVO, science was controlled and its role confined to that of research. Apart from a few scientists who worked on the subject but remained outside AVO, science was thus effectively accommodated.

Zorgen voor morgen on the other hand hinted that the modest eco-modernist regime of Winsemius was not up to the task. Furthermore with *Zorgen voor morgen* scientists partly regained their independence. They now appeared in a more radical coalition together with VROM policy-makers and leaders of the environmental movement. Liberated from the confines of the AVO project, they came up with a contrasting view of the ecological crisis. It illustrates the emergence of a conflict over the meaning of ecological modernization, a conflict which potentially could have great institutional consequences. Interestingly the scientists drew on the same facts but within a different frame. Ironically, here scientists used pathos to regain both their independence and their authority.

5.6. ECOLOGICAL MODERNIZATION AND THE NEGOTIATION OF SOLUTIONS

> The critique that we are moving too slowly is quite right. Yet we cannot speed up.
>
> Minister Winsemius, 1985[160]

One might think that the social consensus on the general goals of Government policy (saving the forests) and the lack of a controversy

distinction of global, continental, fluvial, regional, and local environmental problems. Later he even wrote a book entitled *Gaiasofie* in which he sought to combine Lovelock's theory of global ecology with a New Age philosophy of humanity (cf. Zoeteman, 1989). Zoeteman is presently Director-General at the Ministry for VROM.

[159] See Hajer 1992*a*, 1992*b*.

[160] Minister Winsemius on the issue of acidification, in *De Stem* 26 Oct. 1985.

over the scientific facts (75 per cent emission reductions agreed as necessary), was a strong impetus for radical regulatory action. Yet the paradox of Dutch environmental politics showed in the serious discrepancy between the discourse on required emission reductions and the actual regulatory package. This section focuses on the practices through which the feasibility and acceptability of solutions were defined within the Dutch regime of ecological modernization (i.e. what measures should be taken).

The endorsement of ecological modernization had serious institutional consequences for Dutch regulatory practice which came out particularly strongly in the (re)distribution of legal competences. Until the mid-1980s the actual controlling government agency was not central government but the province. Here individual firms could apply for conditional permits while the provincial government also held the right to take the initiative for a 'revision permit'.[161] The procedure for granting permits included the possibility for third parties to object.[162] In this context Government, industry, and NGOs basically debated what was to be seen as a feasible abatement strategy, be it on highly unequal terms. In general the Dutch Government used the concept 'best practicable means' as a guiding principle. The economic and technical capacity of individual firms thus determined policy outcomes.[163]

The 1970s system was highly unpopular. Environmental NGOs argued that the question should be what nature could endure without fundamental damage, not economic feasibility. The CRMH, the Dutch equivalent to the Royal Commission, had also emphasized the need to come to a more ecologically sound relationship with nature, albeit in very abstract terms. This was hardly surprising given the wide variety of interests that were represented in the CRMH. Although the CRMH was the most prominent advisory

---

[161] The pollution inspectorates were also organized on a provincial and sometimes regional level, but were co-ordinated and supported by the Chief Inspectorate that came under the ministry for VROM. The Inspectie voor de Milieuhygiëne was founded in 1962 but arose from the Staatstoezicht op de Volksgezondheid (Public Health Inspectorate) which dates back to 1865.

[162] The *Wet Algemene Bepalingen Milieuhygiëne* describes the exact procedure for granting permits. Regulation by *Algemene Maatregel van Bestuur* (General Directive, AMVB) is described in the *Wet op de Luchtverontreiniging* (Air Pollution Bill) of 1972 (art. 88).

[163] As Dr Leendert Ginjaar put it, 'we always sought to keep negative environmental effects within acceptable limits. How far can you go: What is technically possible? What is economically feasible?' He continued: 'Now I think it was a different period. A totally different period.' (Dr L. Ginjaar, interview).

body to the government, it did not have the authority to set the agenda of politics. The regulatory practice of the 1970s also worked unsatisfactorily for industry. Firms frequently had to apply for more than one permit at various different authorities and although the national government set the overall goals, regional and provincial inspectors and politicians would often determine whether or not a permit could be handed out. Last but not least, central government more than once had been annoyed with provincial coalitions of left-wing politicians, inspectors, and campaigning groups who aimed for much more stringent regulation than The Hague thought expedient.[164] This was all to change with the advent of ecological modernization.

Verinnerlijking *and the Definition of Feasibility*

Verinnerlijking was described above as a particular Dutch variation of ecological modernization that drew on the long accommodative traditions of Dutch politics. It is also obvious, however, that the *verinnerlijking* approach might thereby also reproduce the inherent limits of the accommodative regulatory regime. The success of *verinnerlijking* depended on co-operation and consensus and such an approach imposes limits to what counts as an acceptable solution. The limits to such an approach are illustrated by Winsemius's thoughts about the workings of *verinnerlijking*:

You try to make them responsible. This also implies that you have the duty to take them seriously . . . It implies that if you do it well, you have to count with their interests. You say if something is going too far. If something is going to be counterproductive. That is also a part of a consensus approach.[165]

In this sense the social limits to problem closure within the Dutch variation on ecological modernization were contained in the very idea of *verinnerlijking*. Winsemius was convinced that 'nobody

---

[164] It is important in this respect to note that the first Lubbers Cabinet that took office in November 1982 explicitly took the free market recommendations of the so-called Wagner Commission (after its chairman, the former president-director of Shell, G. Wagner), as general guideline for its policies. The central concern was to improve the general climate for business. In this context it explicitly called for a simplification and centralization of the permit procedures but also for minor issues like an increase of the maximum sulphur content of fuels. See Adviescommissie inzake het industriebeleid, 1981.

[165] Dr P. Winsemius, interview.

should be asked the impossible'. Basically this meant that you can raise the targets of your policy but you should beware of a target overload. But this of course begs the question how you determine these critical loads, this time not of nature but of industry.

This came out very clearly in the regulation of acid rain. Acidification became one of the themes that had to be led through the policy life cycle. The issue had first been recognized as a serious hazard, partly because of the VROM awareness campaigns. Subsequently effective solutions had to be found. The so-called *Optimalisatiestudie* (Optimalization Study) of July 1984 had shown that in the short run the abatement of SO_2 (rather than NO_x or NH_3) was the most cost-effective option, and so the government decided to focus on the abatement of SO_2 emissions.[166] Consequently the debate became confined to the question of which SO_2 polluter was going to contribute what. Both the refineries and the electricity industry could contribute. At the time it was already apparent that the discrepancy between the actual critical load and the emissions called for a reduction of about 75 per cent. On the other hand ecological modernization included a commitment to find solutions that would not endanger the prerequisite of social consensus and economic prosperity. This was the dilemma facing the regulator.

In the meantime ecological modernization had entailed a huge centralization. Central government had taken back the legal authority to determine the decision-making both on the level of emissions and on the actual technologies that were to be used to achieve a given end. From February 1986 the government could rule via *Algemene maatregel van bestuur* (AMvB, a general directive).[167] This new practice was immediately applied to the regulation of acid rain with the so-called *AMvB Stookinstallaties* (combustion plant directive) of June 1987. The AMvB required full FGD at coal-fired power-stations but left the refineries relatively untouched.[168] Interestingly, although the government argued that it sought to follow as closely as possible the German Large Combustion Plant Directive, it decided to exempt the refineries

---

[166] Although the debate on the regulation of NO_x and NH_3 continued, this section will, for the sake of comparability and feasibility, focus on the negotiation of solutions for the reduction of sulphur dioxide emissions and leave the debate on NO_x and NH_3 reductions undiscussed.

[167] See the new article 20 in *Staatsblad*, No. 655, 12 Dec. 1985.

[168] This intention had first been announced by the government in August 1984.

from the new policy package. The AMvB argued that the SO_2 emissions of refineries had been left aside because of the difficult position of the refineries that operated in a competitive international market. The position of the refineries would be evaluated in December 1988 and December 1993. Then the relationship between emission norms and competitiveness and the relation between emission norms and acidification would be re-examined.

The regional authority of the Rijnmond area wanted to impose far stricter regulations on the refineries but saw their power to do so withdrawn as part of the new regulatory regime. An essential element of the change in legal practice of February 1986 was the restriction on the discretion of the actual permit giving authority.[169] This state of affairs led to fierce criticism, for instance from the CRMH. They warned that it would have a negative effect on the implementation of state-of-the-art technology and urged the government to allow the provinces to set tougher norms if they thought this expedient. In other words, the CRMH thus argued that the system that had replaced the rigidities of the 1970s was less flexible and less amenable to the alleged commitment to innovation of ecological modernization than the system it replaced.

Why were the refineries exempted? And how was feasibility determined? There is no simple answer to these questions. We may start with a formal answer. Formally, feasibility was to be defined according to the principle of best practicable means. That was also reiterated in the AMvB. BPM had been defined by the Minister for VROM in a letter to parliament of January 1984.[170] Old plant would have to install *basismilieuvoorzieningen* or 'basic abatement equipment', while new plant would be required to install state-of-the-art equipment. Yet this included the premiss that this concerned the

latest technological developments of which the feasibility both technically and economically has been shown. The latter condition means, among other things, that the costs are not supposed to be unacceptably high for a normally functioning concern. *Stand der techniek* will in many cases correspond with what is defined as the best technical means for existing sources.

---

[169] See *Staatsblad* 1987, No. 164; Dumas and Keizer 1987.
[170] See *Staatsblad* No. 164, 20; *Handelingen Tweede Kamer*, 1983–4, IMP-Lucht 1984–1988, 18 100, ch. XI, no. 95.

In other words, the Dutch definition largely corresponded with the British understanding of BPM. Technical experts from the refineries[171] and from the ministries of Economic Affairs and VROM as well as with the provincial Inspectorate would discuss the feasibility of abatement options.[172] The structural asymmetry of power in this practice was similar to Britain. First, the governmental experts would always be highly dependent on the explanations of the industrial process from industry's experts. The modern refinery process is simply far too complex for outsiders to understand. Secondly, the market situation of the refineries was just as opaque as the technicalities of the plant. What is more, the government itself was split, since the Ministry of Economic Affairs generally saw its task as to protect industry against any irresponsible claims from other departments.

The situation at the electricity plant, on the other hand, was relatively transparent. FGD was an established technology and unlike refineries that frequently adjust their processes, electricity is one simple process without too many discontinuities. Furthermore, the fact that electricity utilities were state-owned and worked on a monopolistic market made them a soft target.

Surely the technical and economic differences must have contributed to the actual choice of regulation, yet my investigations indicate that there were other more straightforward strategic reasons that left the refineries untouched. The choice of applying FGD at the coal-fired power-plants was a package deal.[173] It was the result of a complex powerplay between the departments of VROM and Economic Affairs. In the early summer of 1984 top officials at VROM found out that Shell and the Department of Economic Affairs had kept relevant information away from VROM.[174]

---

[171] The refineries took one common line; Shell (being the biggest) did the negotiations.

[172] For an excellent study of the political lobbying of Shell and the interaction between government officials and industrial experts, see Barmentlo 1988. However, she missed the fact that these deliberations were often overdetermined by negotiations and trade-offs at the highest level, as we will see shortly.

[173] Dr P. Winsemius, interview.

[174] Shell was about to build a new installation to produce lighter fractions in its refinery. This so-called HYCON process made it possible for Shell to fire its refineries with natural gas which was an easy way to reduce SO_2 emissions. In other words, VROM might have insisted on tougher norms had it known about the installation of the HYCON. Yet the information was kept from VROM and the HYCON was thus initially kept out of the considerations that led to the AMVB.

Because of this the Department of Economic Affairs had secured a very good deal on the new permit for Shell. Since the negotiations over the electricity industry (that also came under EZ) were still open, Winsemius suggested a deal to his colleague at Economic Affairs. VROM would swallow the permit for the HYCON if EZ would agree to the installation of FGD for the electricity plants. Once minister for Economic Affairs Van Aardenne had agreed, they together persuaded the treasury and kept a united front in parliament.[175]

This apparent historical contingency should not obscure the more structural trend. The exemption of the refineries was a recurring element during the 1980s. The *Gigantenregeling* exempted the refineries and other large energy consumers from the increase in prices in the early 1980s; the *Raffinaderijen regeling* exempted the refineries from the general reduction of sulphur content in fuels that was ordained in the *Besluit zwavelgehalte brandstoffen* (1980); and the *AMvB stookinstallaties* secured a continuation of the old practice in the face of a policy commitment to get the acidification problem under control.

One may emphasize the asymmetry of power relations at the level of experts (see above). Yet on the other hand it is obvious that the Government and a multinational firm like Shell also have a relationship of mutual dependency. Regulation is not simply a matter of forcing private parties to comply, it also requires a reasoned trade-off with all kinds of other concerns. Here we find a discursive boundary that constantly needs to be reconfirmed and redefined. One not only wants to know what is effective, one also needs to know how cost-effective action can be secured. The question then is what costs are acceptable, in terms of job losses, economic growth, profit rates, but also the effects for something as vague (and essential) as the general business climate. In such conditions politics becomes a trust game where actors have to compete for credibility. It is remarkable to see how such processes evolve.

What comes out is that in the Netherlands sustainability is discursively subordinated to feasibility. This is the discursive

---

[175] Part of the deal was that EZ would pay for the extra costs of the retrofits (Winsemius, interview). This package deal explains why Winsemius was very proud of the achieved agreement while NGOs and progressive MPs were utterly disappointed.

subordination that dominates Dutch discourse on environmental regulation. Hence the question is how actors come to share understandings of feasibility. Asked to describe the process in which the feasibility of measures at large industrial works is determined, key actors invariably suggested this is the outcome of an intricate process of negotiation. This negotiation practice was not introduced under the new partnership of *verinnerlijking* but was already common practice before ecological modernization restructured environmental discourse. Especially when large firms were involved, the procedure of granting of permits had always been a social negotiation during which well-informed experts from industry and government tried to reach an agreement on what action should be taken. Yet here we will focus on the way in which the more general terms within which that expert debate is to be conducted are set. This implies that we have to deal with the role played by far less knowledgeable actors such as politicians, general managers, or relative outsiders such as campaigners from environmental NGOs who also participated in this debate.

In the regulation of acid rain there seem to have been at least two different practices that helped define solutions that were both feasible and politically acceptable. First and foremost the direct negotiations with polluting firms, but against a background of the communication with representatives from the environmental domain at large. Both practices, were governed by their own set of rules. The examination of these practices and their discursive constructs provides further answers. First we will examine the communication between state and industry, subsequently we will deal with the intercourse between the state and other social groups.

Ecological Modernization at Work: Continuity Comes before Sustainability

In Dutch environmental discourse feasibility had to be defined in the context of the discourse of ecological modernization and sustainability. Within these confines it was the discursive construct of continuity that confined the definition of feasibility. Winsemius described his point of departure as follows:

There are three criteria: Effectivity, Efficiency and Equitability ... Of course, you can order an actor to change his practices all at once. But is that fair? Wouldn't you break his competitiveness? With economic feasibility you have to stay within the boundaries of what a society can take, and what the industries in that society can take.[176]

Obviously, the state of the environment can be such that it requires tougher policies than industries can take. According to this philosophy you then touch upon the limits of legitimate political action. The anticipated reaction then inhibits a tougher policy. As Winsemius said, 'I would like to see the time at which the largest party of the country comes up with a boycott of the largest company.'[177] For industry continuity also had a clear priority over sustainability. Indeed, sustainability is described in such a way as to be logically subordinate: if continuity is at risk, who can pay for the necessary investments? President-director van Duursen of Shell argued:

We aspire to curb all harmful emission on a reasonable time scale. We want to have a leading position in this . . . [yet] I have been arguing for years to the gentlemen from government: 'listen, continuity is what it is all about'. Because if a firm looses its continuity, it no longer has the funds to pay for the abatement installations. . . . Charges, for instance, will always have to be of such nature that a firm does not go broke or is limited in its profitability to such an extent that it can no longer pay for the environmental improvements. [MAH: 'Is this an internal contradiction?'] It is a contradiction but only in time, not in the end result.[178]

Although the shared perception of limits between the former minister for VROM and the President-director of Shell are obvious, these limits to the realm of negotiation have to be intersubjectively defined, adjusted, and constantly reconfirmed. Even preconceived ideas require reconfirmation to have a regulatory effect. It is for that reason that key actors have to keep up their communicative practices to create, reproduce, or indeed challenge the shared perceptions of feasible regulatory measures.

It does not come as a surprise that new communicative practices were opened up with the structuration of ecological modernization as policy discourse. After all, there must have been some concern about the potential repercussions of a more offensive environmental regime. One of the important practices in this respect

[176] Dr P. Winsemius, interview.
[177] Winsemius in *De Stem*, 26 Oct. 1985.
[178] P. van Duursen, interview.

was the informal meetings of the minister of VROM with a number of captains of industry. These so-called 'legs on the table' meetings were proposed by van Duursen, President-director of Shell Nederland. The idea was welcomed by Winsemius. The meetings, that took place about three times a year were described as informal, without an agenda or minutes and were meant to 'resolve strategic issues'.[179]

It was precisely through such top-level exchanges that a discursive order was created that gave meaning to the proclaimed 'new partnership'. The influence of Shell was discursively reconstituted and reinforced in the symbolic order of ecological modernization that defined the terms feasibility, continuity, and acceptability as well as the relationship between these terms. Former minister Nijpels characterized the importance as follows:

Well, it is of the greatest importance to exchange thoughts about issues that are important at that particular moment in time. We discussed general issues, like environmental policy but never details. [MAH: You did not speak about what industry could take?] Well everything was raised: complaints, intentions, ideas of both sides. [MAH: And that formed a basis of your estimation of what was a feasible strategy?] You do not talk not to use the acquired information.[180]

These top-level meetings resulted in the intersubjectively defined trade-off between feasibility and social acceptability which marks the confines in which experts had to come to agreements. It was not a place for threats but for mutual discursive readjustment. Straightforward threats and lobbies focused on a much lower level. At the top exchanges were abstract reconfirmations of largely symbolically defined understandings. However, this should not be interpreted to imply that VROM did not try to act on the SO_2 emissions from the refineries. Yet, once continuity was accepted as limit to regulation and as basis for the new partnership, VROM could not do much better given the asymmetrical distribution of relevant knowledge. It was within this discursive order that Winsemius could make the enigmatic statement that 'The critique that we are moving too slowly is quite right. Yet we cannot speed up.'[181] It epitomized the central paradox of Dutch environmental politics.

[179] P. van Duursen, interview.   [180] E. H. T. M. Nijpels, interview.
[181] Minister Winsemius on the issue of acidification, in *De Stem*, 26 Oct. 1985.

The Politics of Accommodation at Work:
NGOs as Parliamentary Counsellors

The discursive order in which continuity was seen as superior to sustainability formed the basis for the new partnership. Interestingly, despite the general commitment to a sustainable solution to the acid-rain problem, the argument that sustainability should not be allowed to interfere with continuity was hardly ever challenged. In actual fact radical actors mostly challenged the definition of feasibility rather than the implicit order between continuity and sustainability. 'What is feasible? I have always maintained that much more is feasible than industry argues. What is more, one can make timetables for change.'[182] Yet how should this claim be substantiated given the structural asymmetry of knowledge? The most serious challenges were mounted by environmental NGOs like Natuur & Milieu. They made their estimates of what was feasible on the basis of information acquired either through its extensive network of informal contacts in the Ministry, research institutes, or universities or on the basis of findings from literature.[183] Yet this was hardly a serious challenge to the information provided by industry. The most sophisticated attempt to break through the knowledge monopoly of the refineries came in 1986 as two researchers from the University of Groningen published a report suggesting that the abatement of SO_2 emissions by the refineries might be considerably cheaper than had hitherto been suggested. They pointed out that FGD was not the most obvious strategy. Internal adjustment of the combustion process could achieve a reduction of the SO_2 emission by a factor 8 to 10 at a far lower cost.[184] Yet this challenge to the (ideological) barriers of the new partnership faltered. Essential in this respect seems to have been the fact that there was no institutional context in which the refineries could be forced to discuss this intervention. Admittedly, questions were raised in Parliament and an expert from Esso replied to the article (in a personal capacity), but this led

---

[182] Dr R. de Boois, interview.
[183] The journal *Environmental Science and Technology* was particularly important in this respect (J. Fransen, interview).
[184] Pullen and Wiersma 1986.

to nothing more than the reassurance that 'it was not that simple'.[185] It thus exposed the limits to reflexivity of Dutch ecological modernization.

NGOs were not effective in 'pulling' the boundaries of discursive space either. Of course, NGOs argued that the refineries, and especially Shell, had unjustly got off scot-free and persisted in their campaign that labelled Shell as the biggest acidifier. Yet what was required here was a specialist follow up in which NGOs could repeat their claims in substantiated form. However, in the corridors of power of departments and parliament Natuur & Milieu played a totally different role. In parliament they acted as an accommodator, providing MPs with information and preparing motions.[186] Often, their lobbyists knew best what proposals would get through.

I was always pleased when Natuur & Milieu came to the Standing Committee. We discussed things like what do you plug into other parties.... Natuur & Milieu also drew up motions, they were a kind of homing pigeon between the various parties, defining what was feasible.[187]

As such Natuur & Milieu fulfilled an integrating role in the production of the dominant definition of (political) feasibility: 'I became aware you need to avoid motions that do not pass. They are often counterproductive ... So time and again you find yourself balancing: how far should we go, how far *can* we go.'[188] This implies that the NGOs filtered environmental discourse leaving out what one could call 'radical truths'. They thus helped to reproduce the moderate integrative variation of ecological modernization.

We can take as an example the motion of 14 May 1984 that called for complete FGD on power-plants. In this case Natuur & Milieu not only helped MPs to draw up their motion, their work was based on the advice of the top officials from the Ministry for VROM. They were informed that a motion that called for complete FGD application would suit Winsemius best. This advice was

---

[185] Interestingly, recent investment plans (1992) by Shell will introduce roughly the same measures that Pullen and Wiersma suggested in 1986.

[186] Natuur & Milieu would prepare commentaries on all relevant policy documents. These would be a major source of information for environment spokespersons. With regard to the activity on preparing motions Jan Fransen, the air-pollution campaigner, argued, 'I think it is fair to say we initiated the De Boois motion.'

[187] Dr de Boois, interview.   [188] J. Fransen, interview; emphasis added.

taken.[189] This implied that Natuur & Milieu did not work towards a motion calling for action to control the refineries.

The above already indicates that level of accommodation of the Dutch environmental movement. Not only did their air-pollution campaigner Jan Fransen play a role in the drawing up of motions, he also always played an important part in the discussions of the CRMH and influenced MPs. What is more, Natuur & Milieu even had direct access to the Minister. Winsemius argued that the environmental movement should make sure that the ministry 'stayed awake' and had to prevent it from becoming sluggish. For this purpose Winsemius organized informal meetings with the NGO Natuur & Milieu. 'I went to Utrecht every three or four months. We discussed what we thought important. We had agreed that if someone asked whether we had met we would reply "we have agreed that we haven't spoken to one another." '[190] The meetings fulfilled a role in the endorsement of integration and accommodation. Nijpels took this integrative practice one step further. He not only met regularly with the leaders of Natuur & Milieu but became a habitual visitor to regular meetings of both Natuur & Milieu and the more radical NGO Milieudefensie.[191]

The examination of the various communicative practices show that nearly all actors active in the environmental field participated in the reproduction of the dominant environmental discourse. The possibility of arguing with the minister made it attractive to accept his terms in order to increase opportunities to be persuasive, yet it hindered a radical campaign. By 1985 the discourse-coalition around ecological modernization embraced almost all the discursive space in the Dutch debate on acid rain. The exceptions were the (small) radical environmentalist NGOs that refused the label of ecological modernization.

We were dead against ecological modernization. For us it was the Law of Continued Gloom. The examples of Japan and San Francisco were clear enough . . . We were arguing for volume policy. Of course, we demanded FGD, catalytic converters and energy conservation but most important was the reduction of our mobility and energy usage itself. Reduction of

[189] Based on the analysis of information on memos found in the archives of Natuur & Milieu.

[190] The offices of Natuur & Milieu are based in Utrecht; Dr P. Winsemius, interview.

[191] E. H. T. M. Nijpels, interview.

cattle stock has always been our policy. Technological solutions would only imply a further monopolization and increase of scale within the system.[192]

Industry, on the other hand, emphasized the importance of having many communicative channels at its disposal, be it for a different purpose. Van Duursen:

Social values change constantly. We are part of that society. So we have a task to make sure that society keeps an eye open for the restrictions to a new value orientation . . . You have to make sure that you are part of that social process. Such that you can act upon this change.[193]

Practices such as the meetings with NGOs or the CRMH, it may be inferred, are at least as important to integrate the environment in to mainstream concerns and to prevent the environment from becoming a catalyst for a social critique. At the same time one may wonder how effective the discursive inclusion was for the actual closure of environmental problems.

The question is whether discursive practices actually allow for meaningful discussion instead of scanning existing preferences, for instance, through the comparison and discussion of different regulatory scenarios. The research indicates that under the existing practices the debate on the sustainability of abatement options time and again drowned in clichés. Particularly important in that respect was the reference to the concern over the international competitiveness of the refineries. This precluded the normative discussion on what ecological modernization should be about. Winsemius used the competitiveness argument several times in Parliament. 'If measures are taken regarding the refineries this implies an extra burden for these industries that operate already 60% or 70% of their capacity. . . . That extra burden can tip the scales. That cannot be the intention.'[194] And in December 1984: 'We have simply thought it expedient . . . to mitigate the measures. After all, there is a chance that you will chase them away.'[195] Yet

---

[192] T. de la Court, interview.

[193] Later adding: 'Influence, well, that has such a negative connotation. Taking part in the process sounds more positive.' P. van Duursen, interview.

[194] Dr P. Winsemius, *Handelingen Tweede Kamer*, 1983–4, UCV 95, 14 May 1984, 25.

[195] Dr P. Winsemius, *Handelingen Tweede Kamer*, 1984–5, UCV 32, 10 Dec. 1984, 57.

in the interviews the informants readily admitted that phrases like the 'international position of trade' or 'international competitiveness' are completely devoid of meaning. Earlier Winsemius had argued that one really cannot tell to what extent the international position of trade will be affected by certain policy packages: 'It is a mixture of analysis and beliefs.'[196] Looking back, Winsemius argued that although the refineries were in a more difficult position than the electricity industry, 'one should not make too much fuss. The Germans, the great competitors, had to cope with at least the same policy package.'[197]

Even Shell director van Duursen admitted that 'international competitiveness' does not hit the core of the problem. According to him sustainability is much more a matter of long-term investment:

If you want to be sustainable, you have to get away from end-of-pipe technologies. But then you must allow industry five years to develop this technology. You also have to be clear what happens once that period is over and there is no solution. Is that the end or do you get another year?[198]

What was important for the refineries was that they faced 'a level playing-field'. Given the huge investments and the long time horizons involved (an investment in refineries has to be recovered in a ten-year period but the business prospects of an investment are usually fifteen to twenty-five years), industry needed to have a reasonable certainty regarding future regulatory requirements.

The failure to come to more stringent measures regarding the refineries seems to be the result of the failure of parliament to stand by its commitment to prevent the worst damage, coupled with the unwillingness of Winsemius to seriously challenge the representation of the economic impact of the abatement measures on the refineries. After all, this would immediately endanger the new partnership. Here it seems the lack of a credible external force that could have pulled the boundaries of ecological modernization was clearly significant.

It indicates the subtle way in which multinational firms exercised their power. In a global economy shifting investments elsewhere is a very real possibility (although this is arguably less true for a multi-billion oil refinery than a car-manufacturing plant, let

---

[196] Dr P. Winsemius, in *De Stem*, 26 Oct. 1984.
[197] Dr P. Winsemius, interview.　　[198] P. van Duursen, interview.

alone a software house). Yet the threat of withdrawal is not an explicit element of the way in which a multinational secures its interests. Interestingly, it seems that plain threats are very infrequently uttered by the actors who are in a position actually to execute them. The power-effects are in fact the result of a play of positioning between various actors, and are played out through many different practices. The power-effect is the joint product of the fact that MPs mention the possibility of withdrawal in parliamentary debates, the fact that a minister does not deny this although he knows better, the exposé of business and investment figures in the 'legs on the table' meetings between the minister and the president-director, the annual informal tour of MPs through the factory, the newspaper reports on the fragility of Dutch industry, and the difficulty for critical scientists to get answers to their question concerning the set-up of industrial processes. Indeed, the Dutch case also shows how the exercise of the discursive power of a firm like Shell is not confined to trying to influence the minister not to impose too many restrictions. The appreciation of the importance of being part of the process, of keeping open lines of communication, of knowing how debates are evolving,[199] of influencing the conceptualization of new problem fields, but also of associating oneself with certain concerns, illustrates how well aware big actors are of the importance of the discursive realm.

In the end this unsustainable equilibrium was challenged when a coalition of policy-makers, environmentalists, and scientists joined forces and came up with *Zorgen voor morgen*. VROM policy-makers had lost their faith in its technocratic regulatory regime. They realized that by taking the environmental issue apart, and by trying to find feasible regulatory solutions for each issue individually, the total picture was lost. This was not only true of the way in which acid rain was dealt with, it was true of the environmental issue as a whole. At VROM people slowly moved to a more radical position. In the previous section we saw how scientists had become aware of the discursive confines within which they had to operate. Consequently it became expedient to call in the more

---

[199] Striking in this respect was the communication of Thijs de la Court, a fierce opponent of the idea of ecological modernization and certainly one of the most radical environmentalists in the Dutch acid-rain controversy. He mentioned that he had been struck by the fact that wherever he spoke, there always was a representative of Shell present to monitor what he said (T. de la Court, interview).

radical environmental experts, who had often been involved in campaigning and advising since the 1970s. It is in this context that the origin of reports like *Zorgen voor morgen* and the NMP should be understood.

Through these reports VROM sought to open the discursive confines of the Dutch regulatory regime and pave the way for a truly integrative approach. Until then time horizons had, in actual practice, not gone beyond a couple of years and cross-boundary pollution still frequently occurred.[200] A key policy-maker in the Air Division expressed the dilemma of how to deal with challenges to the prevailing regulatory strategy:

In the mid-1980s NGOs argued 'well, that is where we want to go, so let's start.' We as VROM policy-makers knew better: if we really had taken that course, we would have been shot by other ministries, other interest groups and you would have been back to where you started from. So you orient yourself on small steps, feasible steps. . . . Only when we were preparing the NMP did we say 'this is not going to do the job'. Everyone was working in his own little field while thinking he was not going to get big changes through. And so this change will never happen. In the mean time the economy continued to grow. In the autumn of 1989 [he must mean 1988], we as policy-makers decided to go for the big steps anyway. . . . We no longer argued 'it is not feasible' we chose to say 'it has to be feasible'.[201]

The original idea is still evident in the radical goal-setting and the presentation of a picture of the desired environmental quality in 2010. It implied emission reductions of 80 to 90 per cent. It was perfectly obvious that end-of-pipe technologies would not be able to bring about this change. As such it seemed to be the ideal way to launch the big push to fight the environmental crisis. However, here the environmental technocrats reckoned without the host. Their colleagues in other departments and politicians who were not committed to active environmental policies could not obstruct their framing of the problem but were effective in preventing the formulation of a regulatory package that would have matched their presentation of the state of the environment. The NMP, an utter disappointment to many, was, according to Nijpels, 'a reflection of what was politically feasible'.[202] The NMP faltered on

---

[200] Dr B. C. J. Zoeteman, interview.
[201] M. Bovenkerk, interview.   [202] E. H. T. M. Nijpels, interview.

precisely the same obstacle of departmental resistance and interest-group pressure that had led Winsemius to accept in 1985 that in acid-rain abatement 'we cannot speed up'.

5.7. CONCLUSION: APOCALYPTIC DISCOURSE AS ACCOMMODATIVE PRACTICE

From the late 1970s onwards the Dutch Government (and more particularly the Department for VROM) started to push eco-modernist ideas as an alternative to the restricted legalistic policy discourse of the 1970s. The VROM campaign did not meet much resistance. There was a growing awareness of the ineffectiveness of the general rule/permit system even though the system had only been systematically conceived in the 1970s. The acid-rain issue most certainly facilitated this change in the policy discourse. Acid rain not only opened up the established definition of Dutch air pollution but, together with the phenomenon of soil pollution, illuminated the limitations to the top-down legalistic approach to environmental regulation. In the Netherlands 'acid rain' was quickly submerged in 'acidification', which only reinforced the suggestion that problems like acid rain constituted a new generation of structural problems that called for a different regulatory regime. By 1985 the structuration of the discourse of ecological modernization was complete. This case-study shows that ecological modernization had become the only discourse in which actors could credibly express their concerns and priorities. The seriousness of the problem was not contested, but this was combined with the conviction that the problem could be overcome by readjusting the regulatory regime to match the new reality (or newly discovered reality) of structural ecological problems. The previous sections have given ample evidence of the extent to which the discourse of ecological modernization dominated the policy-making at VROM after the mid-1980s. But of course, the dominant position of ecological modernization was not simply the product of the Government's self-monitoring. The change is better understood as the product of a new discourse-coalition that gained influence because many different actors, including industrialists and environmentalists, could employ the new cognitive frame starting from different premisses.

Yet the paradox of Dutch environmental politics consists in the complete mismatch between the generally avowed commitments for effective environmental management and the failure to come to effective problem closure. This paradox is rooted in the fact that the eco-modernist discourse-coalition combined the active promotion of an apocalyptic depiction of the problem with the endorsement of a policy package that fell seriously short of its own goals. This mismatch between meta-apocalyptic story-lines on damage and micro-technical discourses on solutions was the structural characteristic of the Dutch regulation of acid rain. The detailed research of the various practices through which the acid-rain problem was discussed shows how the uniting force of the acid-rain story-line (an example of apocalyptic depiction) concealed various differences in understanding and interpretation. Environmental discourse was in fact rather fragmented, something which did not come out in the apparently superficial discourse on the need for action. Hence, it was in the less obvious practices that the consensus broke down and the many differences in opinion over what the problem really was came out. This research into the micro-physics of power lays bare the subtle structural features of the politics of Dutch environmental discourse.

The seemingly effortless mastery of discursive space by the new eco-modernist policy discourse indicates the first feature typical of the Dutch discursive style. Dutch environmental discourse is predominantly centripetal and socially inclusive. Here it is almost the exact opposite of the British discursive style with its characteristic antagonistic positioning on nearly all aspects of the problem. In the background the political culture of coalition cabinets clearly played a formative role. Yet one might also argue that this was rooted in the religious diversity of Dutch society combined with the great influence of religion on Dutch political life.

The Dutch acid-rain case demonstrates, however, that this does not imply that there are no political disagreements. The accommodative discursive style was just as much a discursive order, with its own rules of prohibition, exclusion, and disciplinization. This case-study shows how the quick endorsement of ecological modernization helped to fulfil the task of social accommodation. Yet one could also argue that the subsequent failure to come to problem closure would lead to a legitimation crisis, and one which the government called upon itself. Here we encounter a second feature of the discursive style of Dutch environmental politics. It consists in the characteristic split between the argumentative interplay on the recognition of an issue and the argumentative interplay on the actual regulation. The discursive space of Dutch environmental politics has two separate chambers: the Chamber of Concern and the Chamber of Regulation. It is a division with long historical roots. It comes out clearly in Table 5.2, which is a representation of the discursive space of the Dutch debate on acid rain. The table illustrates that the general endorsement of the discourse of ecological modernization in fact hid important differences of opinion.

Table 5.2 illuminates a third feature of the Dutch discursive style. The two chambers of Dutch environmental politics each have their own genre. The debate in the Chamber of Concern (cf. 'Status of acid rain as policy issue') is heavily dominated by apocalyptic discourse. Yet the debates in the Chamber of Regulation are determined by feasibility concerns. We saw how in the practices through which the image of damage was created the environment was sacrosanct and occupied a place above everyday concerns. Trees were seen as an indicator of the moral health of society. In the Chamber of Regulation, however, environmental concerns were subordinate to economic considerations. The Dutch acid-rain controversy thus reflects the same separation between the realm of apocalypse and the realm of pragmatic regulatory response that we noticed in the 1972 Urgent Memorandum.

Taken together these three features—the integrative debating format, the discursive division in two chambers, and the dominance of differing concerns (apocalyptic rhetoric and feasibility concerns respectively)—help to explain the paradox of Dutch environmental politics. For instance, Winsemius drew on apocalyptic discourse when he depicted acid rain as an issue of dying forests and crumbling sculptures which created an image of acid rain as a potential threat to our common civilization. The resolute interpellation of Winsemius, right at the beginning of the controversy, most certainly took the wind out of the sails of radical NGOs. Furthermore, the fact that VROM used this apocalyptic presentation as a prelude to the formal introduction of the positive-sum story-line of ecological modernization, only served to enhance the credibility of the eco-modernist discourse-coalition, since it thus also took away the fear from the other side of the political spectrum. In quite a different way the report *Zorgen voor morgen* also drew on apocalyptic discourse to create a radicalized

TABLE 5.2. *The discursive space of the Dutch acid-rain policy debate*

Growth commitment	ECOLOGICAL MODERNIZATION Shared core	Environmental commitment
	Status of acid rain as policy issue	
	– apocalyptic construction: 'I do not want to cause panic but . . .' (Winsemius)	
	– science: 'vitality of Dutch forests is alarming'	
	– creation of doom: 'Yesterday, today, tomorrow?'	
	– acid rain as herald of coming global crisis: 'Slowly the earth is dying and the inconceivable—the end of life itself—becomes thinkable' (Queen Beatrix)	
	– moral discourse, 'dark side of our welfare'	
	– romanticized conception of nature: the innocent, unspoiled	
	– tree as moral indicator	
	Repercussions of acid rain for policy domain	
1988:	– clean environment as pre-condition for economic growth	– new policy discourse needed because of seriousness of new issues
– increase in speed limit is an element of a sustainable strategy		– structural change might be required (reduction of volumes, new modes of transport, reduction of mobility)
– new policy discourse needed because of malfunctioning of 'hierarchical system of command and control'		
– against law of continued gloom: technical solutions can get us a long way		

– managerial discourse, efficiency concerns		– within concern for quality of life – resurrection of the subject – tendency to conceptualize domain in structural terms – politicized discourse, moral concerns
Science and expertise		
– sharp definition of critical loads to find limits	– critical loads as inclusionary device	– critical loads as credible instrument to curb pollution
– strict distinction of science and policy	– reconfirmation of civil legitimacy: CRMH as channel of communication	– problem-oriented engineering approach to scientific research
– high level of generality in tree health surveys		– space-specific analysis of tree health
		– *Zorgen voor morgen*: science as pathos
Confines of environmental regulation		
– ecological modernization as sector policy	– pollution control is as such legitimate	– ecological modernization as innovator of general government policy
– ecological modernization as essentially an efficiency operation		– environmental care to be element of strategy of social change
– solutions to be sought in technical measures	– shared concern over international competitiveness	– structural measures are called for: what is needed is a *Deltaplan* for the environment
– regulation primarily a managerial concern		– regulation primarily a governmental concern
Basic premisses:		*Basic premisses:*
– sustainability < feasibility – feasibility = continuity – nature as economic resource > intrinsic value of nature		– sustainability > continuity – do we really know what is feasible? – nature as ecological resource limits economic exploitation
– 'We cannot speed up'	– 'We are moving too slowly'	– 'More is feasible'

Note: Admittedly, the table omits the radical NGOs that explicitly turned against ecological modernization. Likewise the table does not cover the statements of the MP Willems of the radical PSP (Pacifist Socialist Party) who persisted in framing the issue in terms of 'polluters and polluted'.

version of the integrative eco-modernist discourse. This inter-pellation was not so much an effort to mobilize the numinous authority of science. It was much more based on the idea that having knowledge comes with a social responsibility. The idea behind *Zorgen voor morgen* was clearly also to come to a com-municative coupling of the two Chambers. Science was deployed to correct the feasibility bias in the Chamber of Regulation.

These three features (for all their generality) are not unique to the environmental domain. In actual fact, all three have their counterparts in other domains of Dutch politics and have long historical roots. The integrative debating format is linked to the archetypical image of the Dutch polity as consociational demo-cracy[203] or a case of 'strong' neo-corporatist social arrangements.[204] The Netherlands does not have a tradition of antagonistic politics but of sharing political space.[205] Within that context politics was predominantly conducted by drawing on strategies of selective inclusion.[206] Many historical explanations have been given: the uniting influence of the struggle against water, the absence of an absolutist state, the containment of the working class by the bour-geoisie. However, the consociational arrangements as described by Lijphart and Daalder, foundered on the repercussions of the political, economic, and cultural changes of the 1960s.[207] The secularization and individualization of that time undermined the 'pillarized' social arrangements, and the call for democratization frustrated the consociational settlements. Indeed, after a period of social turmoil that extended from the late 1960s to the early 1980s, a more centrifugal tendency emerged in Dutch politics.[208] Politic-ization and radicalization induced a new sense of identity that subsequently changed the perception of social relations and led to the formation of new social practices.

It was precisely in this period that the new class of environmen-tal issues emerged. The classic explanations of Dutch politics, however, are too general and dated to help to explain the devel-opments in the environmental domain since 1972. What is more,

---

[203] Lijphart 1968, Daalder 1966. For a critique see Stuurman 1983.
[204] See Lembruch 1984.   [205] See Hemerijck 1993.
[206] Selective inclusion, because certain radical actors were kept out, such as the communist trade union and communist party in the post-war period.
[207] See e.g. Middendorp 1979, van den Berg and Molleman 1975.
[208] See Middendorp 1979.

the environmental issue emerged as part of a new agenda of social issues that led precisely to the collapse of these traditional arrangements. More important therefore are the developments in the dynamics in the environmental domain sketched in Chapter 3. Nevertheless, there is an irony in this process. In the Netherlands a decade of radical social action in the 1970s came to a close as the environmental movement turned from radical utopianism to 'chequebook-activism'.[209] As acid rain started to receive widespread public attention from 1982 onwards, the centrifugal moment in Dutch environmental politics was in fact already waning.[210] Partly this was the result of the move towards the centre by the environmental NGOs. Arguably more important was the early recognition of environmental problems by the Government and political parties, both from the right and the centre. The recognition of environmental problems and the quick acceptance that institutional arrangements should be changed, lured others to accept the terms for environmental discourse that were introduced by the government. Unlike in the United Kingdom, the Dutch regulatory regime was supported by practices of co-optation through which NGOs were quickly drawn into advisory discursive practices which came with their own (implicit) rules and restrictions. So within fifteen years accommodative practices were drawn upon again, although in a different place and on a different basis.

Similarly, the characteristic discursive division of the debate on recognition and concern from the debate on regulation has long historical roots. This is closely related to the third feature, the distinct discursive genres of the two chambers. An apocalyptic construction of a policy problem is not at all an unusual discursive strategy in Dutch politics. On the contrary. It can count as one of the traditional strategies of regulation in Dutch politics. It is rooted in the combination of the influence of protestant religion and the proverbial fight of the Dutch against water that has put its mark on Dutch politics ever since the middle ages. At that time, the threats of the sea and rivers were real dangers. And to be sure, in modern days the threat from the sea is still more than a matter of rhetoric. Yet what is important here, is the fact that the apocalyptic construction has over time itself become a core element of the

---

[209] van der Heijden 1992.
[210] Only the possible placement of cruise missiles in the mid-1980s could mobilize a mass movement on a post-material issue.

reality of Dutch politics. Here the threat gives birth to a story-line that subsequently has a political life of its own.

The functional effects of an apocalyptic construction are two-fold. On the one hand it clears the way for the introduction of comprehensive government intervention. On the other hand the historical embeddedness of the apocalyptic construct reifies the functionality of a collective regulatory effort: only if everyone contributes, will action be effective. The fact that even the neo-liberal Lubbers government responded to the environmental problem with the proposal of a National Environment Policy Plan most certainly reflects the legitimacy and embeddedness of what the Dutch call the *Deltaplan* approach to regulation. Yet what the case-study has clearly shown is that the proposal for a regulatory response ('What is needed is a *Deltaplan* for the environment') is as far as apocalyptic rhetoric can get. It then has to surrender to feasibility concerns.

Apocalyptic discourse may be seen as a structural feature of Dutch politics.[211] In one of the few discussions of the theme, Schama points to the diverse ways in which the apocalypse influenced Dutch society in the seventeenth century. He not only notes the omnipresence of the threat of floods a hundred years after the last inundation, but also notices the influence of the sense of fragility in the attitude of the Dutch towards their historical political existence. Ironically, we can discern a great parallel with the reality of contemporary environmental politics. Zahn argues that the alarmist reception given to *Limits to Growth* more than anything indicated the longing for a new religion.[212] Apocalyptic discourse has its roots firmly in the protestant religion and as Zahn rightly observes, Dutch politics is steeped in religion. Basically *Limits* signified the return of the apocalypse in a secularized attire. Similarly, in the acid-rain controversy we find a problem construction dominated by apocalyptic discourse with an almost explicit reference to the nemesis of the consumer society. MPs spoke of the 'Dark side of our welfare' and moralized materialism, while the minister depicted the crisis as a 'flood of reports'.

It all seems to reinforce the enigma of Dutch environmental politics. How can it be that a society so prone to apocalypse still

---

[211] Nevertheless, hardly any attention is paid to this feature in the political scientific literature. The best sources are Schama 1991, Zahn 1991.
[212] Zahn 1991: 43–7.

fails to act? Understanding does not come from exposing the limits to the influence of apocalyptic discourse. Quite the contrary. The point here is that the feature of the apocalyptic construction of reality is not confined to the environmental domain but is constantly drawn upon in the domain of economic politics too. The Dutch discourse of economic politics is pervaded with an anxiety that can compete with the concern over the imminent collapse of the environment. Although the Dutch have the eighth largest economy of the world, there is a constantly reproduced story-line about the fragility of Dutch economic success. And it is precisely in this idea that Dutch well-being is contingent upon continuous hard work (with its obvious discursive affinity with protestant religion) that provides the parallel with Schama's description of Dutch anxiety in its Golden Age. Schama pointed to the prevailing metaphor that Dutch community is set adrift on the 'great historical oceans'.[213] In today's political discourse we again see the metaphorical usage of the fight against the water to warn against a possible economic downfall of the country. The image that is constantly reinforced is that of the Netherlands as a small dependent country in the maelstrom of global competition. A country with an open economy. A country that has to steer carefully on the high seas of the world economy and cannot afford any irrationalities. Indeed, the prevalent cliché of international competitiveness echoes the same sentiment: the continuity of Dutch trade and industry is constantly in jeopardy. This construct is so well embedded that the Government can use it to support a political choice to refrain from making too many demands on leading multinationals like Shell. Nevertheless, the cliché proves to have considerable persuasive power within the existing institutional fora. The apocalypse is the integrative instrument *optima forma*, yet its usage is not confined to the environmental domain.

[213] Schama 1991: 31.

6

Ecological Modernization: Discourse and Institutional Change

6.1. THE 1990S: STILL THE AGE OF ECOLOGICAL MODERNIZATION

The main historical thesis of this book concerns the emergence of ecological modernization as the new dominant policy discourse in the environmental domain. This final chapter concentrates on the examination of the dynamics of the discourse of ecological modernization and addresses the question of what sort of processes of de- and re-institutionalization ecological modernization has so far brought about and is likely to bring about in the years to come. On the basis of that analysis I will then come forward with some suggestions as to how this process of social change, which was shown to have strong sub-political characteristics, might be made the object of public deliberation and democratic social choice.

Ecological modernization began to transform perceptions of environmental problems from the late 1970s onwards. This first became evident at the level of what I have here labelled 'secondary policy-making institutes' such as the OECD, the IUCN, the UNEP, or the UN Commission on Environment and Development (Brundtland Commission). As the case-study chapters have shown, this policy discourse of ecological modernization also gained an increasingly dominant role in individual OECD countries. Although the case-studies focused on regulation in the UK and the Netherlands, one could easily extend the list of countries where environmental policy-making now draws on eco-modernist ideas, concepts, and categories. It is not only the traditional vanguard countries of ecological modernization, such as Germany or Japan, which can be seen to have reconstructed environmental policy-making along the lines described above. This increasingly holds true for countries like Austria, Denmark, Canada, or, to a certain extent, the

United States since the introduction of the 'Win–Win' environmental policies of the Clinton administration.

The new conceptual language of ecological modernization has made a remarkable career. In the early 1980s the general goal of internalization of ecological consequences into the first conceptualization of ideas seemed an unattainable and, given the prevailing social conflicts over socio-economic restructuring at the time, a somewhat naïve ambition. Yet in the early 1990s this internalization has become everyday practice in many sectors of business and industry (although perhaps not to the degree that some might wish). This has not happened because of sheer idealism on the part of the initiating actors, nor because of the strict rules set by the respective public authorities. It is the consequence, I would argue, of the socio-cognitive dynamics of the discourse-coalition that shaped up around eco-modernist story-lines. The discursive power of ecological modernization manifests itself in the degree to which its implicit future scenarios permeate through society and actors reconceptualize their interests and recognize new opportunities and new trouble-spots.

Let me illustrate. The fact that environmental risks potentially challenge some well-established institutions of industrial society is now broadly recognized. This has resulted in a far more cautious attitude towards ecological matters. For instance, both in government and in industry a simple denial of environmental risks has ceased to be a credible option. The many well-publicized cases in which such a straightforward denial later proved to be wrong have simply reset the agenda. Even a vague suspicion of potential hazards now justifies an investigation. This is not just for ethical reasons: ecological awareness is such that the mere suspicion of ecological risk can suffice to break a firm (for instance through spontaneous consumer boycotts). It is not surprising, therefore, that this has not left the organization of enterprise unimpaired. In some sectors firms simply cannot afford to disregard the potential ecological impact of their activities. This development, as Beck has shown, is reinforced by the fact that insurance firms have increasingly become reluctant to cover activities with potentially grave environmental impact (nuclear power was the classic case, genetic engineering is a more recent one). These tendencies all facilitate the recognition of the heuristic value of a more anticipatory style of operation, whether one is active within government or industry.

Consequently, NGOs like Greenpeace have in some cases effect-ively taken up the role of consultants to industry, while within the firm organizational structures have been adjusted to enhance the built-in ecological sensibility.

The effects of ecological modernization can also be shown in other domains. For instance, it is evident from the case-studies that the practices of 'science for policy' have changed. Here we see precisely the sort of discursive contamination that I defined in Chapter 2. It is the changing context of scientific practice which penetrates its institutionalized procedures and changes the local notions of responsibility and good science. The idea that evidence of environmental decline does not imply doom or failure but should be seen as the stepping-stone on the road to innovation, which is central to ecological modernization, leads to the increasing recog-nition of regulatory science as a domain of its own. Regulatory science is no longer positioned as the inferior brother of research science, but becomes a problem-oriented field of research, with its own virtues and functions. Furthermore, we can see evidence of the effects of ecological modernization in terms of the orientation of governmental policy, new legal practices (such as new opera-tionalizations of liability), and new public cognitions and percep-tions. Together these practices in their turn all help to reproduce eco-modernist story-lines. If one then also takes into account the characteristic positive-sum game format of eco-modernist think-ing, it seems fair to predict that ecological modernization will set the tone of environmental policy-making in the years to come. In all this suggests that ecological modernization is not a sort of trend which can enjoy a tremendous popularity but only for a limited amount of time.

As the above indicates, this is not necessarily principally the result of government regulation in the traditional sense but might be the result of the social resonance of eco-modernist principles. The strength of eco-modernist story-lines is that they bring to life a new way of seeing, with new constraints and new opportunities, that is then recognized and interpreted by various actors within the environmental domain, which subsequently leads to all sorts of adjustments in institutional practice.

However, it would be quite wrong to suggest that now that the phase of discourse structuration has been successful, we can sketch the way in which the discourse will settle and become consolidated in the sort of policy-making practices described in Chapter 1. Indeed, one can still sketch quite distinct scenarios for the future effects of the structuration of ecological modernization.[1] Hence if the 1980s were the decade during which ecological modernization conquered the discursive space of environmental politics, the 1990s are likely to become the decade that will determine the historical character of ecological modernization as a political project. After all, the success of ecological modernization did not mean that the institutions of society suddenly collectively decided to take the very same ecological turn and are now marching together in the direction of a green society. It is much more appropriate to see the significance of eco-modernist discourse as generating a process of de- and re-institutionalization, of disembedding and reembedding with, I would argue, as yet uncertain outcomes. We see all sorts of actors employing eco-modernist discourse, but what sort of relations will solidify and what sort of social reality will it produce? This final chapter seeks to show that the perspective sketched in this book contributes to an understanding of this process so far and suggests a particular emphasis for this process of institutional change in the years to come. For that purpose we will first consider the outcomes of the empirical part of this book.

6.2. DISCOURSE MATTERS

This book seeks to show that social constructivism and discourse analysis add essential insights to our understanding of contemporary environmental politics. Central to that argument is that discourse analysis should not be defined in contradistinction to an institutional analysis. In my definition discourse is not synonymous with discussion and discourse analysis is therefore not confined to the analysis of what is being said. Inspired by Foucault, I have sought to bring out the institutional dimension of discourse, considering where things are said, how specific ways of seeing can be structured or embedded in society at the same time as they structure society. I have sought to show how discourse constrains action but also how discourse opens ways to recreate society, how

---

[1] This goes beyond the scope of this book but elsewhere I have suggested four identifiable tendencies: see Hajer, forthcoming.

specific solidified discursive commitments can be dissolved and social change can be brought about. This institutional dimension also comes out in the definition of discourse in Chapter 2. There we defined discourse as a specific ensemble of ideas, concepts, and categorizations that is produced, reproduced, and transformed in a particular set of practices and through which meaning is given to physical and social realities. Discourse analysis is therefore not to be counterposed with institutional analysis, but is rather a different way of looking at institutions that is meant to shed new light on the functioning of those institutions, how power is structured in institutional arrangements, and how political change in such arrangements comes about. The main theoretical thesis of this book is that one can observe how the institutional practices in the environmental domain work according to identifiable policy-discourses that through their story-lines provide the signpost for action within these institutional practices.

The argument so far has been that the developments in environmental politics depend critically on the social construction of environmental problems. To be able to analyse the social dynamics of problem construction we introduced the concept of discourse-coalition that analyses the formations that shape up around certain social constructs. We analysed how specific story-lines gained in influence, who tried to control them, and from what sort of social positions.

This discourse-analytical approach to the investigation of the environmental conflict differs from mainstream analysis primarily in its anti-realist and anti-determinist stand. This means that it does not accept that the ecological conflict is inherent in the physical facts of environmental change, let alone that it would go along with suggestions that this physical crisis immanently produces (or calls for) certain patterns of social change. Alternatively, it suggests that the nature and the outcome of the ecological conflict are dependent on discursive dynamics. This also implies that the extent to which the alleged institutional failure of the existing institutions of industrial society becomes a political problem is open in principle. For instance, even if the experience of environmental destruction was overwhelming, the ecological crisis would not become a political problem either if people were to see the destruction of certain valued environmental features (say loss of amenity or wildlife) or the loss of valued social assets (safe drinking water) as inevitable and could not conceive of a viable alternative; or if certain actors were to be able to convince society that the existing institutional practices could deal with the problem; or if people believed that they would get something better in return (say increased social welfare). From this perspective acid rain is a story-line that, potentially, brings out the institutional dimensions of the ecological problematique. Acid rain was chosen as case-study as a typical example of an emblem in environmental discourse. It was one of the top issues (if not the issue) through which people understood what the environmental crisis was about in the 1980s.

The first important point that came out of this study is the essential role of emblematic issues for the shifts in policy discourse. The case-studies showed the importance of acid rain as an emblem for the general understanding of what environmental problems were about and illustrated its central role in facilitating larger conceptual shifts. This was evident, for instance, in the constant reference of White Papers and reports of advisory councils to acid rain when they were really addressing the way in which to combat the ecological crisis. Theoretically, this underlines the practical value of the distinction of an emblematic level in environmental discourse. Issues like acid rain come and go but what remains is the emblematic level of environmental politics. Emblematic issues thus play a key role in the definition of solutions to the ecological crisis at large. Indeed, the organizational potential of acid rain as a metaphor can hardly be overestimated.

As an emblem acid rain potentially challenged the established understanding of what environmental policy-making should be about. Indirectly, acid rain could thus challenge the institutional arrangements of what Beck has called industrial society. Perhaps the main finding of the case-studies was not the fact that the regulatory mechanisms failed to solve the acid-rain problem, but the illumination of the fact that the environmental domain is full of practices that make environmental problems manageable for the existing structures of industrial society. If we look at the pattern of regulation (see Table 6.1) we can see that, despite the many differences between the two countries, the immediate institutional changes were rather limited and certainly did not match a structural ecologization as defined by Jänicke (see Table 1.1).

Although the acid-rain controversies each had their particular discursive constructions, in the end both countries sought their

TABLE 6.1. The regulation of acid rain in the United Kingdom and the Netherlands, 1972–1989

	United Kingdom	The Netherlands
Discursive closure	*Problem description:* – reductionist scientific discourse (Is there a problem?) *Definition of remedial strategy:* – 'research scientific' discourse as meta-language (Is action likely to have effects?)	*Problem description:* – apocalyptic discourse invoking religious sentiments *Definition of remedial strategy:* – engineering approach ('acid eq.', 'soil module', 'target loads', 'optimalization studies')
Social accommodation	*Politics of containment:* legitimation in terms of traditional socio-political institutions (Royal Society, Alkali Inspectorate): – late acceptance of problem – close consultation with industry – exclusion of NGOs – expert dispute	*Politics of accommodation:* legitimation in terms of a new 'national coalition for the environment': – early acknowledgement of seriousness – co-operation of industry secured through 'self-regulation' – inclusion of NGOs – general trust in government experts (RIVM)
Problem closure	– piecemeal installation of FGD, low NO$_x$ burners; exemption of industrial works – follows EC legislation on catalytic converters and petrol – no volume reductions, no structural change – environment remains sector concern – SO$_2$ levels remain stable while NO$_x$ and VOCs are rising	– nearly full installation of FGD and low NO$_x$ burners; exemption of refineries – pushes EC legislation on tax benefits for 'clean cars' – no volume reductions, no structural change – environment remains sector concern – reduced SO$_2$ levels are more than cancelled out by rise in NO$_x$, NH$_3$, VOCs

solutions in a conceptually rather limited field (see also Tables 4.1 and 5.2). In terms of Jänicke's distinction of strategies of environmental policy (see Section 1.4) both British and Dutch policy-makers relied heavily on options that would qualify as 'elimination': flue gas desulphurization equipment (FGD), catalytic converters, or slurry-processing plants. Anticipatory story-lines were combined with traditional end-of-pipe solutions. While anticipation is one of the endlessly reiterated eco-modernist principles, FGD is the classic example of a remedial strategy. Volume reductions (i.e. curbing traffic or cattle stock), the presentation of a coherent and resolute strategy on the exploitation of the 'fifth fuel' (energy conservation), or innovative transport policies, were either absent or, in the end, failed to materialize or pay off. What is more, despite the confessed eco-modernist emphasis on the need to integrate environmental considerations into the first conceptualization of ideas, in the 1980s environmental policy by and large remained a sectoral concern that stood apart from other areas of governmental policy, both in Britain and the Netherlands.

As it stands the acid-rain case reveals a paradox. While acid rain was in the end generally accepted as a programmatic issue that called for a change of policy strategies, the selected remedial measures failed to give a material form to that new reality. Acid rain was met with rather pragmatic solutions. Was ecological modernization in the end no more than just a set of story-lines, mere window-dressing, while the institutional routines in the end determined the style of regulation? Although such an inference might initially seem attractive it would not do justice to the actual empirical findings.

The Discursive Regulation of Acid Rain

In order to appreciate the discursive dynamics in the environmental domain one should realize that the shift towards ecological modernization should not be compared to a paradigm shift. Ecological modernization is not something that would cause a Gestalt switch, but was an innovative and thus threatening policy-discourse that met with great resistance at the institutional level. Throughout this book, we have given much emphasis to the fact that this resistance should not be seen as an orchestrated attempt to strangle ecological modernization. We showed it to be related to various practices of micro-power, many, often seemingly trivial, mechanisms that influenced the way in which ecological modernization was interpreted so as to make acid rain manageable for the structures of industrial society. I have explicitly identified specific discourse-coalitions in the environmental domain with identifiable social positions and practices. Here we will not reiterate these empirical findings. Instead we will identify the most important discursive mechanisms that interferred with this linkage of eco-modernist problem definitions to more radical processes of institutional change. The point of this exercise is to show the sheer variety of ways in which discourse influences processes of social change and how the reproduction or demise of social structures depends upon the outcome of discursive interaction.

Story-lines.   On the discursive level acid rain had to compete with a whole array of pre-existing figures of speech that were combined with various institutional commitments. These story-lines had become recurring figures of speech or tropes that dominated public understanding and rationalized and naturalized the existing social order. For instance, in Britain acid rain had to find its place in air-pollution discourse which was fully dominated by the concerns about urban air pollution and health effects. Britain had a proud record in air pollution; Britain had prevailing western winds which made acid rain into a foreign problem; Britain had the best scientists working on the issue. These endlessly reiterated figures of speech helped to sustain the legitimacy of given regulatory institutions. These given story-lines set the markers for the space within which acid rain had to prove its alleged urgency. From the point of pure eco-modernist theory the most logical argument would undoubtedly have been to argue that industrial society produces a cocktail of chemicals in the air that cause all sorts of problems. Hence air-pollution regulation should change from being a fire-brigade that goes after certain pollution incidents and start to work towards general pollution prevention instead. Yet politics requires story-lines. Hence the acid-rain story-line was employed to show that the institutional arrangements no longer functioned properly, more precisely, that the effects of air pollution on the natural environment structurally escaped from our attention. We saw how the acid-rain story-line did indeed nicely illuminate the urban and health biases in air-pollution regulation but at the same time it opened the way for policies that related to the acid-rain story-line rather than to the more structural air-pollution problem (the cocktail of chemicals) which eco-modernist theory went on about. In this sense acid rain was an ambivalent story-line that could be used to illustrate a broader institutional problem but could also be reduced to a pragmatic extension of the monitoring network with equipment to measure a handful of acid-rain-related pollutants. What the acid-rain controversy showed is how such story-lines can subsequently themselves become a political reality in their own right and then stand in the way of more reflexive institutional change (cf. 'symbolic politics' and 'black boxing' below).

Disjunction markers. In both case-study countries the policy-making institutions had their own legitimate ways of denying the institutional dimension of the eco-modernist challenge. These so called 'disjunction markers' constituted an essential element in the analysis. In Britain scientific evidence constituted an unavoidable hurdle for any attempt to change environmental policies. Yet as it came out, this commitment referred to a particular sort of science, epitomized by the Royal Society which had, according to many actors in the field, little to do with the reality of regulatory science and was much more based on the procedural rules of research science (Jasanoff) typical of fundamental sciences such as physics. In the Netherlands, on the other hand, the disjunction marker was what I called apocalyptic hymn-singing. Here the denial of the institutional dimension of the problem did not take the form of a call for more knowledge or proof. Quite the contrary, in the Netherlands there did nor seem to be any questioning of the environmental crisis. Yet here the institutional division of a Chamber of Concern and a Chamber of Regulation allowed for the collective hymn-singing to be followed by a negotiation of decision-making in practices in which a more or less traditional industrialist ethics prevailed.

Symbolic politics. In the UK we saw how, once the simple denial of the problem ceased to be a credible option, the installation of flue gas desulphurization equipment was pushed as the appropriate solution. This can be read as another attempt to break down the abstract and all-inclusive quality of the problem according to ecological modernization, this time by the single act of installing scrubbers: a highly concrete solution. Because governmental critics had chosen moderate discursive strategies, e.g. complying with the reductionist understanding of the acid-rain or air-pollution problem, FGD could indeed be put forward as a case of *pars pro toto*. The prevailing construction of the issue as an SO_2–CEGB–FGD-related problem suggested that society could continue its practices without too much disturbance as long as FGD was installed.

Symbolic politics was also evident in other practices. Once ecological awareness had spread, both Britain and the Netherlands took recourse to the publication of comprehensive policy plans. This was of course meant to co-ordinate action, but it can also be identified as a discursive strategy in itself. The policy plans show how in their effort to control the environmental domain, administrators develop complex sets of symbolic icons and rituals that suggested order and control over physical and social developments.[2] It suggested that governments assumed their responsibility and portrayed them as guardians of the environment.[3]

---

[2] See Torgerson's description of the workings of the 'administrative mind' (Torgerson 1990).

[3] The Dutch government has published two comprehensive policy plans since 1989, outlining a strategy to arrive at sustainable development (see *Handelingen Tweede Kamer*, 1988–9, 21 137, Nos. 1–2 and *Handelingen Tweede Kamer*, 1989–90, 21 137, Nos. 20–1). Both plans shared the characteristic combination of an apocalyptic framing of the environmental problem with pragmatic solutions determined by feasibility concerns. The anticipation of regulatory failure was most striking in the case of the problem of acidification. While the official goals of governmental policy remained the prevention of the worst damage (e.g. no forest dieback), the NMP bluntly argued that even the successful implementation of the NMP proposals would rescue only 20% of Dutch forests, thus illustrating the force with which feasibility concerns penetrated the discourse of environmental care. What is more, the NMP failed to come up with proposals for institutional change that could break out of the dead-lock and set in motion an autonomous process of ecological innovation.

The British White Paper *This Common Inheritance* was published in September 1990. The White Paper took the format of a comprehensive policy plan starting off with an outline of the general principles that were to guide policy-making in the environmental domain. These so-called First Principles by and large complied to the eco-modernist format as described in Chap. 1. They included the need (and possibility) to integrate economic growth and environmental quality, the ethical imperative of stewardship, the need for informed public debate, the subscription to the precautionary principle, and the preference for market-oriented strategies including the selective usage of charges (DoE, 1990: 8–16). However, *This Common Inheritance* was noticeably less innovative in terms of discourse institutionalization. As Weale has observed, the emphasis of its proposals lay heavily on procedural changes and technical solutions (see Weale 1992: 124).

The need for sensory experience. A striking finding of the empirical studies concerned the role of meetings and excursions in the process of persuasion. For instance, Lord Marshall, the long-time opponent of the eco-modernist variant of the acid-rain storyline, is said to have been convinced after a meeting with the leader of the Swedish research group. Hence, not reading the scientific papers, but seeing the work that was being done and meeting the author in his own environment was essential for his persuasion. Dutch MPs were shown to have been heavily influenced by an excursion to see dying trees. Symposia and expert meetings fulfilled a similar role. Again, these practices are known not to be the most effective ways of mediating knowledge but can nevertheless be identified as an essential moment in the process of proliferation and utilization of knowledge and, eventually, policy change. Such practices show the essential socio-cognitive aspect of political change. Face-to-face contacts, visits, excursions, and symposia fulfil a key role in the generation of credibility and trust. However, such exercises are of course a highly dubious source of evidence, especially where risk-society problems are concerned, since in such cases the sensory element is even more fundamentally constructed. Hence the practices that one can identify as key to persuasion are at the same time extremely vulnerable to a critique which undercuts their legitimacy.

The discursive creation of macro actors. The case-studies show remarkable examples of instances where the development and employment of policy-relevant knowledge was prevented. The appointment and acceptance of the Royal Society as referee in the British controversy is a case in point. The Royal Society effectively became what Callon and Latour (1981) call a 'macro actor' which was solely responsible for passing judgement on the true state of affairs. It was a clear example of the creation of a sub-political realm where the political choice of what sort of action was required was taken out of the political discussion because of the particular sort of science that was employed. As we saw, the creation of the Royal Society as macro actor prevented the employment of policy-relevant knowledge that was presented by the majority of scientists who had been working on pollution issues.

The social construction of ignorance. The Dutch case illuminated fascinating instances of what is often called the social construction of ignorance. In such cases relevant knowledge is held apart from the discussion. The case of a policy scientist who used the most elementary statistical programme only to find perfect correlations between tree dieback and the (local!) presence of intensive bio-industry could well have been a basis for action. Similarly the model-makers of the RIVM were not allowed to use their model to suggest possible effects of certain changes in emissions but had to confine themselves to mimicking nature.

Black boxing. Another example of discursive politics is what Callon and Latour have called black boxing:

An actor grows with the number of relations he or she can put, as we say, in black boxes. A black box contains that which no longer needs to be reconsidered, those things whose contents have become a matter of indifference. The more elements one can place in black boxes—modes of thought, habits, forces and objects—the broader the construction one can raise.[4]

This is perhaps one of the most fundamental of discursive mechanisms. Making things appear as fixed, natural, or essential is the most effective way of steering away latently opposing forces. Storylines of course almost always lead to black boxing. To become symbolically effective and politically manageable one almost by necessity will have to sacrifice some relevant relationships. Indeed, it seems impossible to explain the communicative miracle described in Chapter 2 without this notion of black boxing. The many different expert scientists involved in the creation of acid rain as a policy problem could all see how their specialized bits of knowledge became part of a wider discourse while their personal awareness of the conditionalities and inherent uncertainties were lost in the intermediary languages that were created. Hence the bureaucratic processing of ignorance could proceed freely.

Positioning and mutual functionalization. In Chapter 2 I discussed the work on positioning theory. The idea is that people (here actors) do not have fixed roles but are constantly being positioned in discursive exchanges. Positioning obviously requires action but at the same time actors are not totally free, that is to say they will have to take account of all sorts of pre-existing understandings, etc. The tropes and story-lines mentioned above

[4] Callon and Latour 1981: 284.

all have their implicit positioning effects. So, for instance, to say that Britain has a proud record is to say that its pollution inspectorate performs well and delegitimizes and belittles the capacity of critics to judge this state of affairs. Such positionings can be understood as specific speech acts: a statement is effectively an act in which people are put in their place. Such positionings are only politically effective to the extent that they are taken up.[5] People can refuse a positioning and even turn it to their own benefit. Yet this refusal of specific positionings was a rather rare phenomenon for various reasons that we discussed in the case-study chapters.

The effect that occurs around positioning and story-lines is what Hildebrandt has called 'mutual functionalization'.[6] This can be seen as the principle that upholds central eco-modernist story-lines such as the suggestion that the environment is a positive-sum game, that environment and development can go hand in hand, or with the idea that environmental investment creates jobs. In all cases one can identify actors who help to sustain such story-lines, be it for different reasons and often with partially different understandings of how such story-lines are to be understood. Actors may accept certain positionings or adhere to certain story-lines because they see it as a functional moment in their overall strategy or general perspective. In the mean time positionings as well as definitions of problems and solutions are accepted and reproduced.

Structured ways of arguing. Each policy domain appears to have a set of historically specific ways of arguing a case. These discursive formats add credibility to the case that is being argued. This does not mean that one cannot introduce new problems in unconventional ways: it only means that discursive defiances come at a cost. In practice this implies that actors initially tend to follow the formats that can count on a certain respectability.

For instance, the Dutch acid-rain story-line had strong apocalyptic overtones that resembled a hyperbolic representation of reality.[7] This was a format that was shown to have recurred within the confines of the Dutch environmental domain since 1971, but which had much deeper historical roots. Exaggeration was used and recognized as the proven means to create the social support for regulation. In actual fact such a discursive format had grave

---

[5] For very insightful and witty examples, see Davies and Harré 1990.
[6] Hildebrandt *et al.* 1994.   [7] A phrase used by Cuperus (1992: 452).

institutional implications. The apocalyptic construct also suggested that the problem was so massive that it could only be handled in a particular way: a new *Deltaplan* was needed. This was a state-led approach which would almost guarantee full social support. The Dutch acid-rain controversy shows how as a consequence obvious relationships of industrial or agro-industrial sources and environmental effects were obscured (most clearly the relationship between ammonia emission and tree dieback). As has been shown in Chapter 5, the Dutch policy discourse was determined by a discursive affinity between the apocalyptic format, with its deep historical roots both in Christianity and the Dutch struggle against the water, and, on the other hand, the system-theoretical policy format that drew heavily on engineering discourse. This engineering vocabulary became a convenient intermediary language that allowed quarrels to take place over interim goals, emission ceilings, target loads, choices of technical solutions, and the relative optimum contributions of target groups rather than over the actions that could be taken immediately.

Hence the hyperbolic representation created social support but at the same time obscured the fact that acidification was in many respects a rather intelligible problem with, in technical terms at least, several rather straightforward solutions. The mixture of apocalyptic and system-theoretical discursive formulations hindered an open debate on the different possible scenarios under consideration, their effectivity, and their social repercussions. Both discourses in fact served to sustain the social basis for the environmental actions designed and executed by experts.

The British case illustrates similar problems. At first glance the British acid-rain controversy indicates a lively discussion on the meaning of acid rain and the actions that should be taken (see Table 4.1). Yet just as the Netherlands had its two Chambers of Discourse, Britain had various, often rather subtle selective mechanisms operating within the Science-based Policy Approach that at key moments prevented debate from taking place (such as at the meeting at Chequers or the Tenth Report of the Royal Commission). The British controversy was likewise heavily influenced by specific structured ways of arguing according to which people respected the way one talks on this sort of occasion. Whereas the Dutch government actively helped to produce the hyperbolization of the acid-rain problem, the British government stood by its ideology of empiricism. This ideology was in fact so strong that it prompted an anticipated reaction on the part of the eco-modernists and instead of insisting on validating their case by their own epistemological criteria they fell back on traditional empiricist categories and lost their case.

This list of discursive mechanisms is not meant to be exhaustive but it gives the most important mechanisms that could be identified in the British and Dutch case-studies. The discursive mechanisms that we distinguished above may give a practical explanation for why environmental discourse so often combines eco-modernist philosophies with solutions that do not match this eco-modernist format without running into political difficulties. It shows, first of all, the degree to which discourse contains structures that can be as effective in resisting political change as walls and barbed wire can in preventing trespassing. Discursive features such as disjunction markers or structured ways of arguing are essential attributes of policy domains. Not recognizing these discursive structures leads to unduly optimistic and in fact rather technocratic thinking about policy change. Second, this section illuminates the discrepancy between the discursive reality created in an academic intellectual debate on environmental policy and that of an institutionalized policy domain. To understand the fate of intellectual constructs such as ecological modernization one has to recognize how specific symbols and metaphors are structurally embedded in a particular domain. They represent all sorts of institutional commitments and fulfil an essential role in putting intellectual policy theories to the test. Third, the discursive features described above may differ rather dramatically between countries. Hence, although it is fair to say that ecological modernization is an international policy-discourse one really has to allow for national particularities if one wants to come to an assessment of the effects of this discourse on societal processes of de- and re-institutionalization.

The analysis shows that argumentative activity is an independent layer of power practices that is far more complex in its logic than a more conventional realist analyses suggest. The politics of discourse is not about expressing power-resources in language but is about the actual creation of structures and fields of action by means of story-lines, positioning, and the selective employment of comprehensive discursive systems (such as law or physics), etc. In this process actors are by no means completely autonomous:

they are constrained not only by conventional understandings and agreed-upon rules of the game but also by mutual positioning, existing institutionalized routines, and changing contexts. What the analysis of discourse coalitions hoped to show was how these practices can mutually influence and reinforce one another.

Discourse analysis furthermore illuminated the many instances where the immanent institutional challenge was defused. The environmental domain includes a large set of concealed practices that between them managed the institutional challenge. For some, this is potentially a great relief. Apparently the status quo is protected by a multitude of practices that can be seen to uphold the structures of society and avoid a politicization of certain existing arrangements. Yet one may wonder whether this should be a reason for contentment. One could also argue that the prevailing political institutions are unable to find the way in which society can readjust to a new era in which the public voices its concern over the costs of progress through periodic outcries over emblematic issues. If one were to recognize that these issues are symbols, are metaphors that refer to a more fundamental unease, then perhaps the effectiveness of the prevailing institutions in defusing controversies over issues like acid rain as policy problems works to the detriment of the strength of the political sphere in the long term.

Running parallel with the process of defining environmental problems and their solutions is a whole agenda of social change that is effectively avoided by the prevailing policy-making practices. This was shown to be the case for the acid-rain controversy but seems to surface again in the case of what can certainly count as the emblems of the 1990s: the global environmental problems like ozone depletion, biodiversity, and global warming. Below we will briefly discuss these issues and then turn to potential institutional remedies.

The Emblems of the 1990s: The Global Environment

One of the prime findings of this book might be the instrumental role of emblematic issues in processes of conceptual and institutional change. Above we noted how all sorts of discursive practices helped to make the emblematic issue of acid rain manageable for the structures of industrial society in the 1980s. One may wonder whether similar processes now occur with the global environmental problems that dominate the agenda of environmental politics in the 1990s. Do global environmental problems change our perception of the ecological dilemma and the institutional change that is called for?

Once more, realists and social constructivists are likely to provide different answers. Seen from a realist point of view the globalization of ecological discourse might appear as consequence of institutional learning. The more we know about the environment, the more we become aware of the global interconnectedness of environmental problems. Consequently, institutional restructuration may seem only natural and appropriate: now we have come to recognize the global dimension of the ecological crisis we should restructure our policy-making institutions so as to be able to meet this global ecological challenge. Here a social constructivist analysis might come to different conclusions. This would be based on an exploration of the discursive construction of these problems, the discourse-coalition that upholds them, and the potential political consequences of this new emblem in environmental politics. Here we can only hint at the sort of questions that are considered in more detail elsewhere.[8]

The story-lines on global environmental problems owe their sustained central political role to an unlikely discourse-coalition. It embraces, for instance, powerful supra-national environmental NGOs (like the IUCN, WWF, or Greenpeace), most national governments, and the media, as well as some powerful scientific organizations (such as the World Meteorological Organization); at the same time it can count on a great interest from the public (at least in the OECD countries). The story-lines on global environmental problems have already generated an almost unprecedented degree of institutional change. They solidified in influential new policy institutions such as the Intergovernmental Panel on Climate Change (IPCC), dominated the Earth Summit at Rio, are central to Agenda 21, and indeed, they may be said to have caused the creation of a new global policy domain.

The construction of global environmental problems, and global warming in particular, is by no means uncontested. Here one may distinguish at least three critical responses. First, there is a group

[8] See e.g. Benton and Redclift 1994; Wynne *et al.* 1995.

of scientists working on issues such as global warming who argue that problems like global warming are beset with scientific uncertainties. These scientists are best known for having been used by the former Bush Administration in an early attempt to obstruct joint declarations of intent on the reduction of greenhouse gas emissions. Their arguments were quickly swept aside as being a fig-leaf for mere politics although it was—given the inordinate complexity of the phenomenon—of course quite hard to prove they were wrong.

A second source of criticism is the Third World platforms that quite early on took a critical stand towards the globalization of environmental discourse. They saw their own orientations and concerns in environmental and developmental issues being squashed by the rise of the new global problems. What they pointed out was that the global problems had typically been produced mainly by the North, while the solutions seem to have to come primarily from the South. What is more, as authors like Sachs or Shiva have shown convincingly, this Northern bias came out clearly in the proposed integrative remedial strategies, where the further integration in the global industrial complex was set out as being the way forward.[9]

A third important strand of criticism comes primarily out of the social constructivist tradition in the sociology of science. It criticizes the alleged indeterminacy in scientific knowledge-claims *vis-à-vis* their evidential basis. Sociologists of science like Wynne, Shackley, or Kwa have shown that the approach of the working groups of the Intergovernmental Panel on Climate Change favours a particular sort of scientific approach that unnecessarily leads to a centralization of knowledge, an unnecessary reduction of flexibility regarding the inclusion of new evidence, and effectively prevents the application of the knowledge acquired for the development and assessment of various policy scenarios.[10] Their analysis suggests that the scientific consensus around the issue of global warming is the product of the specific way in which this problem is constructed by an identifiable set of actors, institutions, and technologies. Working in a similar vein, Liberatore has illuminated that the interaction of scientific research and policy-oriented

[9] See various contributions in Sachs 1992.
[10] Wynne 1994; Shackley *et al.* 1993; Kwa 1994.

funding constructs global environmental problems in such a way that they allow for certain definable technical solutions or congenial political interventions.[11]

Now, one may of course argue that the fact that the phenomenon of global warming is circumscribed by uncertainty is hardly a ground for complacency. Are there not some criteria of plausibility that we could draw on, or is there not something like a no-regret scenario according to which we should do whatever we can so as to at least prevent possible catastrophes? The point here is that the dictum 'if it does not do good it does not do harm either' does not hold. Of course, a radical reduction of energy levels in the OECD countries seems a sensible strategy (which is, incidently, one of the reasons why environmental NGOs have jumped on the issue of global warming: it seemed a window of opportunity to enhance the credibility of what they had always been arguing). The uneasiness comes from the fact that the case of global warming seems to indicate that the latest constructions of the environmental problematique reinforce a techno-corporatist tendency where policy-making practices aim to control and subsequently solve a specific set of pre-defined problems rather than leaving space for competing problem definitions and rival scenarios of resolution. Although a notion like sustainable development can broadly meet with sympathy we can observe how some fixed regulatory regimes seek to close themselves off from criticism and set about to institutionalize their own particular programme of pre-conceived solutions under that very heading. The question is whether alternative patterns of institutionalization can be conceived that would do more justice to the many different concerns of the participants in this coalition around eco-modernist story-lines like sustainable development and would be able to keep a debate going on the meaning of ecological modernization in specific cases. The remainder of this chapter is devoted to that question.

6.3. ECOLOGICAL MODERNIZATION AND INSTITUTIONAL REFLEXIVITY

In Chapter 1 we introduced the risk society theory of Ulrich Beck as a perspective that might help to come to a critical assessment

---

[11] Liberatore 1994. See also Buttel and Taylor 1992.

of ecological modernization. Beck's risk-society thesis suggests that the complex conglomerate of problems that we lump together under the heading of the ecological crisis show the structural deficit of the institutions of industrial society. According to his analysis, industrial society digs its own grave, since its institutions increasingly show their inability to handle the dangers it itself produced.

I have sought to give the Becksian thesis a discourse-theoretical twist. In my terms the ecological crisis is then a 'discourse of self-confrontation' that calls for a reconsideration of the institutional practices that brought it about. In distinction to Beck I have also defined reflexivity as an essentially discursive quality. Reflexivity is a quality of discursive practices that illuminates the effect of certain social and cognitive systems of classification and categorization on our perception of reality. What does this mean for the future of ecological modernization? This study has shown that an understanding of pollution and environmental degradation as the negative effects of certain systems of classification and categorization is in fact widespread, not only in public practices, but in certain policy practices too. At the same time the break with existing practices remains a cause for concern. Not simply because not enough is being done. The problem is rather the lack of debate on what an ecological modernization should mean. From the social-constructivist perspective of this study we will of course not come up with a suggestion as to how specific pre-conceived goals can be achieved. The challenge does not concern the goal but the process. The challenge seems to be to think of an organization of ecological modernization as a process that allows for social change to take place democratically and in a way that stimulates the creation of an—at least partially—shared vision of the future.

The Ideal of a Reflexive Ecological Modernization

The ideal of a reflexive ecological modernization is a democratic process of deliberate social choice out of alternative scenarios of development (or indeed non-development). This social choice is not confined to instrumental rationality (concerning questions of 'how to do'); reflexive ecological modernization should also stimulate the debate on norms and values that should be the carriers of the modernization processes.[12] From a social constructivist point of view this means that reflexivity should be related to Douglas's definition of pollution as 'matter out of place': reflexive ecological modernization focuses the discussion on the social order in terms of which we define what constitutes pollution.[13] In this model ecological modernization automatically ceases to be a primarily techno-administrative affair in which the objective reality of expert discourse determines what is out of place and where solutions are selected that respect the implicit social order of expert discourse. Alternatively, ecological modernization fosters a public domain where social realities and social preferences determine which actions should be taken, which social practices are to be respected, and which conventions or practices should be changed.

Here a reflexive ecological modernization stands face to face with the familiar techno-corporatist regime. A techno-corporatist ecological modernization above all seeks to find one universal language to facilitate the search for the most effective and most efficient solutions to unequivocal problems. For this purpose it ordains the creation of new expert organizations where the best people can work in relative quiet. The challenge for reflexive ecological modernization lies much more in finding new institutional arrangements in which different discourses (and concerns) can be meaningfully and productively related to one another, in finding ways to correct the prevailing bias towards economization and scientification, and in active intersubjective development of trust, acceptability, and credibility (see below).

A reflexive organization of science and expertise in the environmental domain has already received much attention. It would require the creation of practices that are more open to those corrective influences that were consciously kept away under the dominant ideology of specialization. Interdisciplinary work is as such a simple example of what a more reflexive form of science could be. More fundamentally, the demystification of science should have repercussions for science's monopoly on knowledge claims. A reflexive approach to science would start from the recognition of the conditionality of knowledge, and would also acknowledge the important role of the process in which scientific insights are

---

[12] A central argument from Habermas 1981.   [13] See Sect. 1.2.

processed into workable units of knowledge. To put it in the words of Foucault, a reflexive application of science would start from the realization that science does not produce truth but only truth claims.[14]

Reflexive ecological modernization emphasizes the importance of the mobilization of independent opinions versus the respected power of authorities. In Enlightenment thought science broke through the authoritarianism of religion, but authors like Beck argue that in today's context the scientific rationality and expert authority have themselves become the authoritarian forces that hinder the progress of enlightenment.[15] The institutional practices of privileged expert advice have led to a negation of all sorts of critical capacities in society and have falsely resulted in a delegation of decision-making on some of the most important decisions to experts' councils that operate beyond the realm of democratic control.[16] This has by no means improved the quality of decision-making. What is called for are institutional practices that allow for the playing off against one another of different sorts of knowledge. Rather than orienting ourselves on science as the universal discourse, one might choose to facilitate the institutionalization of a public language that would allow for productive inter-discursive debates. Here scenarios of societal modernization might become the point of integration.

In this model reflexive ecological modernization would seek a strengthening of public, inter-discursive forms of debate in order to contextualize expert opinion and in order to make environmental politics a matter of deliberate and negotiated social choice for certain scenarios of societal modernization. The assumption is that public understanding might well be much more mature than many

[14] As Giddens put it, it would start from the recognition that no knowledge is any longer 'knowledge in the "old" sense, where "to know" is to be certain.' (Giddens, 1990: 40). Beck's analysis of the disciplinary effects of expert discourses is roughly in line with Foucault's general concerns. Yet Beck insists on the possibility of creating liberating practices based on reason and mutual respect whereas Foucault would argue that these practices will themselves become the new cages within which marginal discourses will be destroyed: see Foucault 1975.

[15] Beck 1991: 115. Elsewhere he captured this development by arguing that science had shifted from being a breaker to a powerful creator of taboos: see Beck 1986: 257.

[16] Important examples here are drawn from the control of DNA technology and the development of critical levels of acceptable risk, e.g. in the context of nuclear power or harmful chemical substances.

politicians and experts would like us to believe. Of course, people have shown themselves to be susceptible to apocalyptic story-lines, to manipulation of statistics or of images of damage. But the old pair of the ignorant 'layman' and the omniscient 'expert' no longer encompass reality. Experts have lost much of their numinous legitimacy over the last two decades because of their handling of issues like nuclear power or indeed acid rain. The outright denial of the existence of serious problems has only served to increase the likelihood of a resurrection of the sceptical and autonomously judging citizen. What is more, the era of the new environmental conflict has shown us many instances where 'experts' were produced through the exclusion of many other relevant sorts of knowledge (including complete scientific traditions). A critical public debate could therefore be seen as a clarifying force, contrary to what conservative politicians sometimes suggest, who see public debates as a spectacle in which the man of the street is manipulated by shrewd rhetoricians. Reflexive ecological modernization would require the institutionalization of forms of public debate that could illuminate hidden assumptions and implicit commitments and would, last but not least, generate and proliferate insights into what I have here called the ecological dilemma. After all, the perspective of this book suggests that environmental politics cannot be about solving the environmental crisis: it is about finding patterns of modernization that at best are less wasteful and without unwanted social side-effects. Reflexive debate should clarify options, and clarify costs, and would thus put an end to mediocre naturalist environmentalism and call for more explicit political exposition and reflection on the sort of development society really wants.

The assumption underlying the above is that an extension of the possibilities for open deliberation would help find the socially acceptable strategies of modernization that at the same time might produce better results in terms of problem closure. Strategies that the prevailing institutional arrangements did not manage to develop in the context of the first fifteen years of eco-modernist discourse.

Reflexivity in the Risk Society?

A reflexive ecological modernization as suggested above is of course not without its problems. In the above we have put much emphasis on the virtues of more open debate and democratic deliberations, while at the same time underscoring the vices of unnecessary secrecy and élitist and techno-corporatist tendencies of all sorts. In fact such a plea is more appropriate in some contexts than in others. For instance, it is a more typically European argument to argue that more democratic institutional arrangements would produce better or more desirable outcomes in environmental politics. To the extent that the normatively desirable goal of democratization is facilitated such regulation is undeniably superior. Yet some American authors working in the field of environmental regulation would be likely to respond less enthusiastically were they to discover that the concept of reflexive institutions was synonymous with more democratic arrangements and extended legal rights. After all, some of the arrangements that are currently being promoted in Western Europe as new directions have been in place in the USA for years; without, however, producing policies of a less symbolic sort. Neither did it necessarily bring results that scored better in terms of reduction of pollution levels, the time-frames within which levels were brought down, or the degree to which those involved have been able to reach a consensus on the route to follow.[17] In Europe a Freedom of Information Act, Right to Know schemes, or extended legal opportunities to force political changes through litigation may seem revolutionary ideas. From an American perspective one is more likely to emphasize that such practices are extraordinarily time-consuming, expensive, and not necessarily more effective.[18] And, if argumentative discourse analysis teaches us anything, it is that the format in which policy discourses are developed has a immense influence on the construction of policy problems and the outcome of the political process. Hence it could be argued that such argumentative institutional structures are likely to produce dissensus where consensus might have been achieved in other circumstances.

This concern about the almost automatic conflation of reflexive institutions with more public debates is in fact further reinforced by recent developments in the German and Dutch discussions on environmental issues. The recognition of the role of public debates in the allocation of credibility has led to the employment of what

[17] David Vogel's (1986) comparison between the USA and the UK still remains the classic statement which itself elaborates on the pioneering study by Hawkins (1984).
[18] See Jasanoff 1986; Amy 1987.

is rapidly developing into a full new branch of professionals: the so-called info-brokers. Info-brokers are professional agencies that broach and take up an active role in debates on the part of specific industries.[19] For instance, recently info-brokers initiated a series of debates on behalf of the German chemical industry on the alleg- edly damaging effects of certain chemical works. These debates were open discussions in which reputable politicians, well-known activists, as well as representatives of industry took part. Without the burden of having to come to some sort of regulation these debates ended in an increased confusion and a derangement of the environmentalist critique. They only seemingly enhanced reflexiv- ity by giving an overview of the different perspectives one might have on the chemical works and by creating the possibility for these different groups to meet.

The above can serve as an introduction to a disturbing dilemma that has so far received relatively little attention. This concerns the question of what reflexivity and reflexive institutional arrange- ments could mean if one takes seriously the self-description of modern society as a risk society. In a recent article Keller and Poferl comment on the proliferation of environmental mediation practices in Germany.[20] They observe that the new ecological awareness has intensified the need to legitimize all sorts of prag- matic solution to fundamental environmental problems, such as the siting of waste recinerators, the planning of nuclear waste disposal sites, etc. Here environmental mediation has been discov- ered as the promising institutional practice to meet the new de- mand for legitimation that is an inherent part of ecological modernization.[21] Keller and Poferl provocatively wonder whether such practices do not simply have to falter in the face of recent insights from the sociology of technology and risk, such as Perrow's work on 'normal accidents' or Beck's risk society analysis. After all, there it is argued that catastrophes are an essential and normal by-product of certain industrial processes, that such mishaps can- not be predicted very well, that both the provision of insurance and compensation typically fail, and that ecological problems are increasingly hard to isolate in terms of time and space. These characteristics are at odds with the clearly structured situation

---

[19] I rely here on van der Loo 1992.   [20] Keller and Poferl 1994.
[21] The concept and practice of environmental mediation have been described in detail in Amy 1987.

that is required for efficient environmental mediation. Hence, it is easy to see how institutional arrangements like environmental mediation meet the new demand for legitimation. Yet it is also obvious that mediation sets clear limits to any reflexivity in terms of problem definition.

However, the danger here is one of throwing the baby out with the bath-water. As the example of environmental mediation already indicates, the potential problems seem to be related to specific institutional designs, not to the idea of public deliberation as such. For instance, litigation might be a case which generates many disadvantages, but is to a large extent due to the fact that it locates an exchange of views at the very end of the creation of new plans. The general idea of reflexive institutional arrangements would rather suggest locating debate at a much earlier phase in the decision-making process. Many of the frustrations of governments and firms with participation or legal rights in fact stem from the fact that the existing institutional practices only allow for comments and objections to be made once the plans are almost in their final form. Hence, every second spent on legal cases or participation or evaluation in say environmental impact assessments is lost. On the other hand, all creativity of citizen groups, NGOs, or indeed rival construction firms (as in the case of a protest against governmental schemes for major infrastructural works) is similarly lost. Reflexive institutional arrangements would seek to overcome such negative features of existing institutional designs. Of course, even a maximum of creativity will not simply erase the relevance of the self-description of society as a risk society. But that is hardly a reason to refrain from thinking about reflexive institutional arrangements.

Towards Reflexive Institutional Arrangements

The term 'reflexive institutional arrangements' refers to formats that facilitate deliberate decision-making in the age of the risk society. The social sciences are just beginning to face the task of designing such institutional arrangements.[22] What is more, the above suggests that although new institutionalized forms of public debate may contribute to a 'demonopolization' of knowledge and

---

[22] For some first attempts, see Burns and Ueberhorst 1988; Zillessen 1993.

decision-making, public practices may also have some perverse effects. First of all, if one opts for the maximalization of openness one may create practices that lock actors into certain antagonistic discursive formats. Secondly, if one opts for the more confined and consensual setting of environmental mediation, where a limited group deliberates on what decisions to take, one may have to delimit the definition of the problem to allow the practice to work. In those cases one would miss precisely the bigger issues of environmental and development that should be opened up for discussion.

Argumentative discourse-theory suggests that reflexive institutional arrangements cannot be thought out of the context of a continuous struggle for discursive hegemony. In Chapter 2 we suggested that politics is a process of the creation of discourse-coalitions based on a shared definition of reality. We suggested that credibility, acceptability, and trust determine the extent to which this process of world-making is successful. This implies, first of all, that if one seeks to design reflexive institutional arrangements one should take into consideration the socio-cognitive basis of discourse-coalitions. For instance, the fact that Third World platforms refute the new construct of global environmental problems seems not so much due to a scientific doubt about the importance of global threats. It is more likely that it was the result of the complete lack of trust on their part towards supra-national institutions such as the World Bank that were given a central role in the implementation of Agenda 21.

Secondly, a reflexive arrangement should take account of the social-cognitive factor in the development of its knowledge base. William Dunn argued in his analysis of policy reforms:

knowledge production and use are symbolic or communicative actions involving two or more parties who reciprocally affect the acceptance and rejection of knowledge claims through argument and persuasion. Thus, knowledge is not exchanged, translated, or transferred; it is transacted by negotiating the truth, relevance, and cogency of knowledge claims.[23]

Reflexive institutional arrangements can therefore never be based on pre-conceived problem definitions. Indeed, reflexive practices should in large part be oriented towards constructing the social problem.

[23] Dunn 1993: 265–6.

Third, in the development of reflexive institutional arrangements one should pay due attention to the location of the authority to define the problem, who are the participants, and what sort of solutions can be considered. The design of reflexive institutional arrangements is of course a subject for another book. Yet to give an idea of the sort of arrangements I have in mind I will conclude this book with a presentation of two examples that might guide further thinking: societal inquiry and discursive law.

6.4. REFLEXIVE INSTITUTIONAL ARRANGEMENTS: TWO SUGGESTIONS

The idea of societal inquiry is to create a new platform for debate on issues of modernization, such as major infrastructural works (high-speed rail links or the reorganization of mobility) or emblematic environmental problems like acid rain. A societal inquiry would be a practice where all sorts of sub-political decisions could surface and where different scenarios of modernization might be developed and compared with actual policy strategies. At present governments come up with proposals and society has a chance to voice its opinions afterwards. By its very structure such a format enhances antagonisms and minimizes reflexive potential. The idea of societal inquiry is to provide a public site where this discussion can take place, parallel to the traditional sequence of decision-making. Depending on the issue at hand, societal inquiry might be held on the local or regional level (as for instance in case of a debate on traffic management strategies and local air pollution in and around a specific city), the national level (as with mobility schemes and discussion over the tracks for a fast train, etc.), or even the international level (as with acid rain). Hence the idea is not to enhance critical discourse by providing practices within firms or organizations. The idea is rather to create a civic stage, as the public domain where people contribute knowledge and take part in deliberations principally as citizens (although it is obviously essential that they bring their specific knowledge into the discussion) and which governments, firms, and citizens should come to understand as an essential part of decision-making on issues with great social repercussions.

A concretization of societal inquiry could be as follows. A societal inquiry would take place at public request if an organization or individual manages to get a certain number of citizens to sign up for an inquiry, for instance on the issue of acidification. The societal inquiry would be co-ordinated by an authoritative independent agency that together with the various parties that are involved in a given subject, e.g. acid rain, would select a governing committee for a specific societal inquiry. It would invite people to deliver information and points of view, hear witnesses, gather information and evidence on its own initiative. On that basis it would seek to produce a set of distinct scenarios for development that would then be brought back into the discussion. One of the key problems with the institutionalization of such a new inquiry practice is to define the terms of debate. The study has amply shown how the definition of issues mobilizes political bias. In this respect the ideal character of the societal inquiry can be defined by comparing it with two classic examples from the environmental literature: the British Windscale Inquiry and the Canadian Berger Inquiry.

In his *Rationality and Ritual* Brian Wynne has shown how the semi-judicial format of the Windscale Inquiry (1977) forced the debate on the application of new technologies and assessment of risks in a certain direction.[24] The inquiry procedures defined the controversy to be about technical facts concerning the empirical safety and other direct impacts of the specific plant that was proposed. The scope of the inquiry thus essentially became confined to a debate dominated by experts about technical aspects of the installation, who had the simple task of determining whether the project should go ahead or not. Value conflicts were not recognized as legitimate, nor were concerns about political and institutional control over the new technologies. The Windscale Inquiry thus had strong exclusionary and disciplinary rules that allowed the formulation of some positions but not others. As Wynne argued, these reflected existing, taken-for-granted institutional structures. Reflexive debate was thus pre-empted.

The contrasting example is presented by the Canadian Mackenzie Valley Pipeline Inquiry headed by Mr Justice Thomas R. Berger (1974–7).[25] The Berger Inquiry was in fact a social and environmental impact assessment that origined in the proposals for a new pipeline. The inquiry consisted of several different stages

---

[24] Wynne 1982.   [25] See Gamble 1978.

including informal hearings and meetings with the inhabitants of specific native settlements in the vicinity of the proposed route. It was explicitly seen by its initiators as a forum to 'state, create and develop' the positions of the affected people. Dryzek reports that in the end the Berger commission 'constructed a community position on land, economy, and governance' that constituted a coherent challenge to the prevailing idea of development.[26]

As such the Berger Inquiry is an example of a discursive practice that helped to create a valuable argumentative interplay. The Inquiry was not directly oriented towards policy-making but towards argumentation and debate. This is an essential characteristic for a societal inquiry. Whereas the Windscale Inquiry was confined to an assessment of whether risks were acceptable, the Berger Inquiry made the normative basis of (regional) development into the focus of debate. It sought to create an exchange of ideas concerning the social aspirations of a community and resulted in a negotiated compromise with the original development plans.

An important reflexive quality of such inquiries is that they develop individual concerns into public positions. That is to say, it creates the forum where private concerns can be understood in terms of larger social relations and developments. Ideally a societal inquiry should result in discussion of different negotiated scenarios for development (or non-development). These scenarios should be discussed and compared in various ways.

Whether or not one wants to give a societal inquiry a semi-judicial form is open for consideration. Above we discussed the possible negative effects of such a format. The advantage could be that one creates a setting with high symbolic appeal. It would allow for different parties to present their cases, call witnesses, and question the arguments presented by specific firms and administrative agencies. It would present new possibilities to participants who now have difficulty, perhaps not so much in being heard, as in getting into an argumentative interplay. A societal inquiry would have to be a practice that allows for meaningful debate, that would make it possible to assess the potential alternatives that are proposed by independent experts, NGOs, or local groups, or that are reported from elsewhere in terms of their consequences for various

---

[26] Dryzek 1990*b*: 107; see also Dryzek 1990*a*.

aspects including regional development, social justice, and environmental effects.

This discursive practice would illuminate the cultural and political choices implicit in the technicalities and show what other courses of action might be taken. It could make clear what the agenda of environmental NGOs is really about, just as it would require firms to substantiate what they mean when they talk about the social responsibility of the firm. On the other hand, it would also help firms or policy-makers to explain what their dilemma is. If the choice really were laying off workers or caring for the environment, public debate might take a different turn than the easy care of environmental quality *per se*.

The effects could be various. First, a societal inquiry thus conceived might help to strengthen the appeal of ecological modernization as a common political project for societal change. Ecological modernization could become more of a social event which is not imposed on people but is perceived as a project in which people can recognize a role for themselves and can actively take part. This social component could be further reinforced if societal inquiries employed different scenarios for regulation that would illuminate the choice of social arrangements that is inherent in resolving environmental problems.[27] Societal inquiries would provide a platform for inter-discursive communication in the sense that it would inevitably require participants to engage in debates with actors who have a different understanding of what the problem is. Yet because the participants have an object in common (i.e. acidification) they would have to compete and persuade people as to why their view of reality is a superior vantage-point.

Secondly, a societal inquiry could bring dynamics back into the policy process and correct the issue–attention cycle. For instance, a societal inquiry could also bring acid rain back into the discussion. It could be based on the question 'Why did we not achieve our goals?' The debate would presumably result in the recognition of acid rain as an emblem and would result in new problem definitions and new ideas about what would constitute socially acceptable solutions. This would subsequently filter into new possible

[27] In this case the scenario-method would not take the prevailing form of comparing doing nothing, doing something, and doing everything. If scenarios are truly to reflect choices in societal modernization they should differ precisely in what they aim to achieve and in the way in which they hope to achieve these specific effects.

scenarios for regulation and renewed social pressure. The societal inquiry could be conceived so as to give politicians an incentive to support one specific scenario. The regulation of environmental conflicts would thus break with the naturalistic tendency: it would become clear that there are political choices to be made upon which people have to reflect.

Thirdly, a societal inquiry would correct the balance of power, most notably between firms, government, and NGOs and local groups. NGOs could campaign for a societal inquiry to be held on a particular issue and this would allow them to challenge both firms and governments in a meaningful and highly visible context.[28] What is more, it would allow them to do so on their own terms. Fourthly, societal inquiries could indicate where social consensus is likely to be possible, or where it cannot be achieved and choices will have to be made.

Presumably the discursive practice of societal inquiry would change the debate by creating new cognitions and new societal coalitions. But it is unlikely that they themselves will always produce consensus. Yet this is not necessarily a problem. The important result should be a certain clarity of orientation in terms of the trajectory of modernization in the future. To maximize the impact of a societal inquiry they should therefore aim to produce a set of normative commitments that are widely shared by the participants as well as to make clear where there is fundamental dissensus. This could be combined with the formation of clearly distinct scenarios of development. If politicians were subsequently to commit themselves to different scenarios the argumentative interplay would flow over into the parliamentary debate and would create an opportunity for real political choice on the part of the electorate.

Discursive Law

The idea that guides this search for an enhanced reflexivity is a break with a linear understanding of the policy process. Rather than separating problem definition and policy implementation, one should let these processes run in parallel. This also comes out in what the Dutch legal theorist Willem Witteveen has called 'sym-

---

[28] Weale presents a similar argument in his discussion of the practice of environmental impact assessment: see Weale 1992: 172.

bolic law' but I prefer to call discursive law. This is defined as akin to constitutive legislation (like the American Constitution):

an act or body of acts (or enactments), done by agents (such as rulers, magistrates or other representative persons), and designed (purpose) to serve as a motivational ground (scene) of subsequent actions, it being thus an instrument (agency) for the shaping of human relations.[29]

The interesting idea of this alternative legal theory is that the strength of discursive law is primarily related to the societal debate from which it is derived. The idea is that discourse-coalitions (Witteveen uses Fish's concept of 'interpretive communities') would form around specific examples of discursive law that would keep up a continuous argumentative debate on the meaning and implications of the legal text. This discourse-coalition would produce, reproduce, and transform a set of legitimate argumentations with which concrete problems should be decided upon. The strength of the law would then be determined by the extent to which one succeeds in extending one specific way of conceptualizing reality across the full width of the discursive space.

Characteristic of discursive law is its emphasis on the provision of normative arguments, rather than on the practicalities of specific regulatory cases. Here it could take the arguments that have been developed in the course of the societal inquiry, or even one normative scenario that has come out as the most acceptable and promising course of development, as basis for the legal text. Subsequent regulatory action would then have to elaborate on that societal debate. The discursive legal text will give the judicial argumentations that are seen as legitimate and will thus help actors to determine which actions will be perceived as complying with the goals of ecological modernization and which fail in this respect. These kinds of legal arrangements can only function if actors engage in discussions on the way in which others interpret a text. Following Witteveen I would argue that governmental strategic considerations should be on trial in these continued societal debates. Government thus becomes a matter of defending the operationalization it has given to the normative commitments that were produced in societal debates.

The two examples given here might enhance the opportunity for

---

[29] K. Burke quoted in Witteveen 1991: 76.

people to make a conscious choice regarding the social order in which they would want to live and to which they feel committed. Highly visible practices such as a societal inquiry could well strengthen the social power of an eco-modernist discourse-coalition. The clear presence of a societal debate might enhance people's ability to perceive their own specific problems and struggles in terms of alternative interpretive frames and help create a vibrant project of ecological modernization. This might make an argumentative ecological modernization also more effective in achieving its own goals than its techno-corporatist counterpart.

However, the environmental dilemma of modern society will not go away. Modern society will constantly have to renegotiate what behaviour is tolerable and what is not, what is ecologically and socially feasible and what is not. Yet it would be a great improvement if ecological politics could shed its prevailing techno-corporatist format and create open structures to determine what sort of nature and society we really want.

Appendix: List of Interviews

UNITED KINGDOM

Mr C. Agren, former head of the Swedish NGO secretariat on acid rain (written information), 28 June 1990.

Ms S. Booth, Department of the Environment, co-author of DoE Environmental Policy Review, 24 September 1990.

Dr A. Crane, CEGB Policy Division, 19 June 1990.

Dr P. Freer-Smith, head of site studies, Forestry Commission Research Station, Alice Holt Lodge, 18 June 1990.

Dr R. Grove-White, former chairman of the Council for the Conservation of Rural England, 20 June 1990.

Dr M. Holdgate, former chief-scientist at the DoE, 13 September 1989.

Dr G. Howells, former CEGB scientist, 15 May 1990.

Professor F. Last, former chairman of the Natural Environment Research Council's Committee on the effects of air pollution, and former assistant director of the Institute for Terrestrial Ecology, Pennicuik, 7 May 1990.

Lord Marshall of Goring, former chairman of the CEGB (1982–90), 21 September 1990.

Professor T. O'Riordan, University of East Anglia, former chairman of UK working group on World Conservation Strategy, 28 February 1990.

Dr R. Pearce, scientist, Forestry Institute, Oxford, 30 January 1990.

Mr C. Rose, former air-pollution campaigner with Friends of the Earth, 26 February 1990.

Sir Hugh Rossi, chairman of the Environment Select Committee of the House of Commons, 14 June 1990.

Sir Richard Southwood, former chairman of the Royal Commission on Environmental Pollution, 13 October 1989.

Mr M. F. Tunicliffe, clean air inspector, Her Majesty's Inspectorate for Pollution, 24 September 1990.

Rt. Hon. William Waldegrave, former Minister for the Environment at the DoE, 14 October 1992.

NETHERLANDS

Dr R. M. van Aalst, scientist, RIVM, 19 June 1991.

Ir. J. C. A. M. Bervaes, deputy-director De Dorschkamp, SBB research centre, Wageningen, 14 August 1991.

Ir. W. M. J. den Boer, former official at Forestry Division, Ministry of Agriculture and Fisheries, 1 October 1991.

Dr R. de Boois, former vice-chairman of the Standing Environment Commission, Second Chamber, 2 September 1991.

Ir. M. Bovenkerk, former co-ordinator of acidification policy, VROM, Air Division, 4 September 1991.

Prof. N. van Breemen, Faculty of Soil Sciences, Agricultural University Wageningen, 25 September 1991.

Ir. A. H. M. Bresser, Air Division RIVM, former secretary to Project Group AVO research, 19 June 1991.

Mr T. de la Court, WISE, 21 June 1991.

Ir. P. van Duursen, former president-director Shell-Nederland B. V., 26 September 1991.

Ir. J. T. J. Fransen, air-pollution campaigner with Natuur & Milieu, 5 September 1991.

Dr L. Ginjaar, minister at VoMil 1977–81, chairman CRMH, 11 September 1991.

Mr E. Nijpels, minister at VROM 1986–9, 9 January 1992.

Dr T. Schneider, director AVO research, RIVM, 21 August 1991.

Ir. T. F. C. Smits, Forest and Landscape Division, Ministry of Agriculture and Fisheries, 20 June 1991.

Ir. G. van Tol, Forest and Landscape Division, Ministry of Agriculture and Fisheries, 20 June 1991.

Dr P. Winsemius, minister at VROM 1982–6, 12 September 1991.

Dr B. C. J. Zoeteman, former director, Air Division, VROM, 4 September 1991.

Mr S. Zwerver, former head of the Air Quality Section of VROM, 21 August 1991.

References

PRIMARY SOURCES

Adema, A. H. and van Ham, J. (1984) (eds.), *Zure Regen: Oorzaken, Effecten en Beleid—Proceedings van het symposium te 's-Hertogenbosch 1983*, Pudoc, Wageningen.

Adviescommissie inzake het industriebeleid (1981), *Een Nieuw industrieel Elan*, Ministerie van Economische Zaken, The Hague.

Anon. (1980), *The World Conservation Strategy: Living Resource Conservation for Sustainable Development*, MCN/UNEP/WWF, Geneva.

Anon. (1982), *The Global 2000 Report to the President*, Penguin, Harmondsworth.

Anon. (1983a), *The Conservation and Development Programme for the UK: A Response to the World Conservation Strategy*, Kogan Page, London.

Anon. (1984), *De vitaliteit van het Nederlandse bos in 1984*, Staatsbosbeheer, Utrecht.

Anon. (1985), *De vitaliteit van het Nederlandse bos 3*, Staatsbosbeheer, Utrecht.

Anon. (1986a), *De vitaliteit van het Nederlandse bos 4*, Staatsbosbeheer, Utrecht.

Anon. (1987), *De vitaliteit van het Nederlandse bos 5*, Staatsbosbeheer, Utrecht.

Anon. (1988), *De vitaliteit van het Nederlandse bos 6*, Staatsbosbeheer, Utrecht.

Beament, J., *et al.* (1984), *The Ecological Effects of Deposited Sulphur and Nitrogen Compounds, Proceedings of a Royal Society Discussion Meeting*, London, 5–7 Sept. 1983, The Royal Society, London.

Boer, W. M. J. den, and Bastiaans, H. (1984), *Verzuring door atmosferische depositie—vegetatie*, Publikatiereeks Milieubeheer, VROM, The Hague.

Boogaard, F. and Meens, T. (1985), 'Verzuring Gaat Nog Pijn Doen' (interview with Winsemius), *De Stem*, 26 Oct. 1985, p. W2.

CEGB (Central Electricity Generating Board) (1959), *First Report and Accounts*, HMSO, London.

—— (1983), *Statement by Sir Walter Marshall at Launch of Acid Rain Research Project*, 5 Sept. 1983, CEGB, London.

Central Policy Planning Unit (DoE) (1985), *Environment Policy Review*, Department of the Environment (internal document).

CRMH (Centrale Raad voor de Milieuhygiëne) (1981), *Advies over The World Conservation Strategy*, Staatsuitgeverij, The Hague.

—— (1983), *Milieu van jaar tot jaar 1982*, Staatsuitgeverij, The Hague.

Chester, P. F. (1989), *The Large Combustion Plant Directive: How Effective will it be?*, paper presented at Conference on Strategic Planning for the Control of Emissions from Combustion Processes—The Latest Information, CEGB Archives.

Commission of the European Communities (1983), *Third Action Plan for the Environment*, Office for Official Publications of the European Communities, Luxemburg.

Court, T. de la, *et al.* (1987), *Zure Regen—Gisteren. Vandaag. Morgen?*, WISE, Amsterdam.

DoE (Department of the Environment) (1975), *Controlling Pollution—A Review of Government Action Related to the Recommendations by the Royal Commission on Environmental Pollution*, Pollution Paper No. 4, HMSO, London.

—— (1976), *Effects of Airborne Sulphur Compounds on Forests and Freshwaters*, Pollution Paper No. 7, HMSO, London.

—— (1976), *Pollution Control in Great Britain: How it Works—A Review of Legislative and Administrative Procedures*, Pollution Paper No. 9, HMSO, London.

—— (1977), *Environmental Standards: A Description of United Kingdom Practice*, Pollution Paper No. 11, HMSO, London.

—— (1979), *The United Kingdom Environment 1979: Progress of Pollution Control*, Pollution Paper No. 16, HMSO, London.

—— (1982), *Air Pollution Control—The Government's Response to the Fifth Report of the Royal Commission on Environmental Pollution*, Pollution Paper No. 18, HMSO, London.

—— (1984a), *Acid Deposition—Reply by Patrick Jenkin (Press Notice)*, 22 Mar. 1984, DoE, London.

—— (1984b), *Controlling Pollution: Principles and Prospects—The Government's Response to the Tenth Report of the Royal Commission on Environmental Pollution*, Pollution Paper No. 22, HMSO, London.

—— (1984c), *Statements at Munich Conference*, printed in House of Commons Environment Committee 1984 (Minutes of Evidence, DoE), 299–307.

—— (1984d), *Acid Rain: The Government's Reply to the Fourth Report from the Environment Committee*, HMSO, London.

—— (1986), *Conservation and Development: The British Approach*, HMSO, London.

—— (1988a), *Our Common Future—A Perspective by the United Kingdom on the Report of the World Commission on Environment*, HMSO, London.

DoE (Department of the Environment) (1988*b*), *Air Pollution: The Government's Reply to the First Report from the Environment Committee*, Session 1987–8, HMSO, London.

—— (1989), *Digest of Environmental Protection and Water Statistics*, HMSO, London.

—— (1990), *This Common Inheritance—Britain's Environmental Strategy*, HMSO, London.

—— (1991), *This Common Inheritance—The First Year Report*, HMSO, London.

—— (1992), *Digests of Statistics*, HMSO, London.

Departement van VoMil (1979), *SO$_2$ Policy Framework Plan*, pub. 1981, Departement van Volksgezondheid en Milieuhygiëne, Leidschendam.

Departement van VROM (1984*a*), *Verzuring door atmosferische depositie* (IWACO-rapport), Publicatiereeks Milieubeheer, no. 2, VROM, Leidschendam.

—— (1984*b*), *Environmental Program of the Netherlands 1985–1989 (IMP-environmental management 1985–1989)*, VROM, Leidschendam.

—— (1987*a*), *Milieuprogramma 1988–1991 Voortgangsrapportage*, VROM, Leidschendam.

—— (1987*b*), *Tussentijdse evaluatie verzuringsbeleid*, VROM, Leidschendam.

—— (1987*c*), *Leidraad Besluit Stookinstallaties*, Publikatiereeks Lucht, No. 67, Ministerie van VROM, The Hague.

—— (1990), *Comparison of Environmental Policy Planning in Industrial Countries in the Context of the National Environmental Policy Plan*, VROM, The Hague.

Dumas, M. T. and Keizer, V. G. (1987), *Leidraad besluit stookinstallaties*, Publikatiereeks Lucht, VROM, The Hague.

Ecologist, The (1972), *Blueprint for Survival*, Penguin, Harmondsworth.

Feenstra, J. F. (1982), *Cultuurgoederen and luchtverontreiniging*, VROM-reeks Lucht, No. 1, SDU, The Hague.

Forestry Commission (1985), *Report on Forest Research 1985*, HMSO, London.

—— (1986), *Report on Forest Research 1986*, HMSO, London.

—— (1987), *Report on Forest Research 1987*, HMSO, London.

—— (1988), *Report on Forest Research 1988*, HMSO, London.

Fowler, D., Cape, J. N., Leith, I. D., Paterson, I. S., Kinnaird, J. W. and Nicholson. I. A. (1982), 'Rainfall Acidity in Northern Britain', *Nature*, 297, 3 June, 383–6.

Friends of the Earth (1988), *Windsor Great Park Beech Health Survey*, FoE, London.

Handelingen Tweede Kamer (1971–2), Urgentienota Milieuhygiëne, 11 906, no. 1–2.

—— (1982–3*a*), IMP Lucht 1981–1985, 16 495, no. 1–2.

Handelingen Tweede Kamer (1982–3*b*), IMP-Lucht 1981–1985, UVC 17, 21 Feb. 1983, 25–42.

—— (1983–4*a*), De problematiek van de verzuring, 18 225, no. 1–2.

—— (1983–4*b*), De problematiek van de verzuring, UVC 95, 14 May 1984, 1–44.

—— (1983–4*c*), IMP-Lucht 1984–1988, 18 100, no. 1–2.

—— (1983–4*d*), IMP-Lucht 1984–1988, UCV 41, 19 Dec. 1983, 19–43.

—— (1983–4*e*), Onderzoeksnotitie, 18 224, no. 18.

—— (1983–4*f*), Meer dan de som der delen, 18 292, no. 1–2.

—— (1984–5*a*), Indicatief Meerjaren Programma Milieubeheer 1985–1989, 18 602, no. 1–2.

—— (1984–5*b*), IMP-Lucht, UCV 32, 10 Dec. 1984, 41–66.

—— (1987–8), Regeringsstandpunt over het rapport van de World Commission on Environment and Development, 20 298, no. 1–8.

—— (1988–9), Nationaal Milieubeleidsplan—kiezen of verliezen, 21 137, no. 1–2.

——(1989–90), Nationaal Milieubeleidsplan Plus, 21 137, no. 20–1.

House of Commons Environment Committee (1984), *Fourth Report: Acid Rain*, (2 vols., including Minutes of Evidence), session 1983–4, HMSO, London.

—— (1987), *Historical Buildings and Ancient Monuments*, i, session 1986–7, HMSO, London.

—— (1988), *Air Pollution*, (2vols., including Minutes of Evidence), HMSO, London.

House of Lords Select Committee on the European Communities (1982), *EEC Environment Policy: The Proposed Third Action Programme*, 16th Report, session 1981–2, HMSO, London.

—— (1984), *Air Pollution*, 22nd Report, session 1982–3, HMSO, London.

—— (1988), *Alternative Energy Sources*, 16th Report, session 1987–8, HMSO, London.

Independent Commission on Disarmament and Security Issues (Palme Commission, UN) (1982), *Common Security: A Programme for Disarmament*, Pan, London.

Independent Commission on International Development Issues (Brandt Commission, UN) (1980), *North South: A Programme for Survival*, Pan, London.

—— (1983), *Common Crisis*, Pan, London.

Innes, J. L. (1987), *Air Pollution and Forestry*, Forestry Commissions Bulletin No. 70, HMSO, London.

—— *et al.* (1986), *Forest Health and Air Pollution: 1986 Survey*, Forestry Commission Research and Development Paper 150, Forestry Commission, London.

Interfutures (1979) (ed.), *Facing the Futures*, OECD, Paris.

Johnson, B. (1983), *The Conservation and Development Programme for the UK—A Response to the* World Conservation Strategy, Kogan Page, London.

Langeweg, F. (1988) (ed.), *Zorgen voor Morgen: Nationale milieu-verkenning 1985–2010*, Samson/H. D. Tjeenk Willink, Alphen aan den Rijn.

Marshall, W. (1983), *Launch of Acid Rain Research Project Royal Society and Scandinavian Academies—Statement by Sir Walter Marshall, FRS*, Lord Marshall Archives.

—— (1986), *Chairman's Statement on FGD—11 September 1986*, Lord Marshall Archives.

Mason, J. (1990), *The Rationale, Design and Management of the Surface Waters Acidification Programme*, mimeo.

Meadows, D. L., Randers, J., and Behrens, W. W. (1972), *The Limits to Growth—A Report for The Club of Rome's Project on the Predicament of Mankind*, 1978 edn., Pan, London.

NOS-Journaal (1983), *Transcript Eight O'Clock News*, 25 Feb. 1983, NOS, Hilversum.

OECD (1977), *The OECD Programme on Long Range Transport of Air Pollutants: Measurements and Findings*, OECD, Paris.

—— (1979), *Declaration Meeting of Environment Ministers*, OECD, Paris.

—— (1980), *Environmental Policies for the 1980s*, OECD, Paris.

—— (1981a), *The Environment—Challenges for the '80s*, OECD, Paris.

—— (1981b), *The Costs and Benefits of Sulphur Oxide Control: A Methodological Study*, OECD, Paris.

—— (1984), *Future Directions for Environmental Policies in Environment and Economics*, OECD, Paris.

—— (1985), *Environment and Economics—Results of the International Conference on Environment and Economics*, OECD, Paris.

Ooyen, D. van, and Court, T. de la (1984), *Het zure regen boek*, Vereniging Milieudefensie/WISE, Amsterdam.

Pulles, T. and Wiersma, D. (1985), *Luchtverontreiniging door Olie-raffinaderijen in een veranderende Markt*, IVEM Report No. 9, IVEM, Groningen.

Reed, L. E. (1976), 'The Long-range Transport of Air Pollutants', *Ambio*, 5, special issue on the International Conference on the Effects of Acid Precipitation, Telemark, Norway, 14–16 June, 164–8.

Rose, C. and Neville, M. (1985), *Tree Dieback Survey, Final Report*, FoE, London.

Rossi, H. (1984), *Open Letter to Sir Walter Marshall*, 14 Nov. 1984, Committee Office HoC, London.

Royal Commission on Environmental Pollution (1971), *First Report*, HMSO, London.

—— (1974), *4th Report: Pollution Control: Progress and Problems*, HMSO, London.

Royal Commission on Environmental Pollution (1976), *5th Report: Air Pollution Control: An Integrated Approach*, HMSO, London.

—— (1983), *9th Report: Lead in the Environment*, HMSO, London.

—— (1984), *10th Report: Tackling Pollution—Experience and Prospects*, HMSO, London.

Schneider, T. and Bresser, A. H. M. (1987), *Dutch Priority Programme on Acidification*, Verzuringsonderzoek Eerste Fase, Tussentijdse Evaluatie, Report No. 00–04, RIVM, Bilthoven.

—— —— (1988), *Dutch Priority Programme on Acidification—Evaluatierapport Verzuring*, Report No. 00–06, RIVM, Bilthoven.

—— and Heij, G. J. (1991) (eds.), *Review Report Dutch Priority Programme on Acidification*, Report No. 200–8, RIVM, Bilthoven.

Schumacher, E. F. (1973), *Small is Beautiful: A Study of Economics as if People Mattered*, 1974 edn., Abacus, London.

Schütst, P. *et al.* (1983), *So stirbt der Wald*, BLV, Munich.

Scottish Development Department (1989), *Acidification in Scotland 1988—Proceedings of a Symposium*, Edinburgh, 8 Nov. 1988, Scottish Development Department, Edinburgh.

Scottish Wildlife Trust (1985), *Report of the Acid Rain Inquiry*, proceedings of the meeting held in Edinburgh, 27–9 Sept. 1985, Scottish Wildlife Trust, Edinburgh.

Smits, T. F. C. (1991) (ed.), *De vitaliteit van het Nederlandse bos 9*, Ministerie van Landbouw, Natuurbeheer en Visserij, Utrecht.

Stichting Maatschappij en Onderneming (1973), *Tussentijds bestek—vraagstukken rond milieu en economische groei*, SMO, The Hague.

Stuurgroep Verzuringsonderzoek (1985), *Additioneel programma verzuringsonderzoek*, VROM, The Hague.

Tangena, B. M. (1984), *Optimale bestrijding verzurende emissies*, Publikatiereeks Lucht, No. 40, VROM, Leidschendam.

Torrens, I. M. (1984), 'What Goes Up Must Come Down: The Acid Rain Problem', *OECD Observer*, No. 129, July.

United Kingdom Centre for Economic and Environmental Development (UK CEED)/National Economic Research Associated (NERA) (1986), *Environmentalism Today—The Challenge for Business, Proceedings of a One-Day Conference*, London, 16 Apr., UK CEED, London.

UK Review Group on Acid Rain (1982), *Acidity of Rainfall in the United Kingdom—A Preliminary Report*, Warren Spring Laboratory, Stevenage.

—— (1983), *Acid Deposition in the United Kingdom*, Warren Spring Laboratory, Stevenage.

UNEP (1983), *The State of the World Environment 1983*, UNEP, Nairobi.

Waldegrave, W. (1986*a*), 'Economic Development and Environmental Care—The Role of Government', in UK CEED/NERA, 1986, 5–12.

—— (1986*b*), 'The Implications of the UK Government's Policy on Acid Rain', *Proceedings of the Joint Conference on Acid Rain—The Political*

Challenge, 19 Sept. 1986, Institue for European Environmental Policy/National Society for Clean Air, London, 7–14.

Ward, B. and Dubos, R. (1972), *Only One Earth: The Care and Maintenance of a Small Planet*, Penguin, Harmondsworth.

Warren Spring Laboratory (1972), *National Survey of Air Pollution 1961–1971*, Warren Spring Laboratory, Stevenage.

Watt Committee (1982), *The European Energy Scene*, Report No. 11, Watt Committee, London.

—— (1984), *Acid Rain*, Report No. 14, Watt Committee, London.

—— (1988), *Air Pollution, Acid Rain, and the Environment*, Report No. 18, Watt Committee, London.

WCED (World Commission on Environment and Development) (1987), *Our Common Future*, Oxford University Press, Oxford.

Winsemius, P. (1986), *Gast in eigen huis—Beschouwingen over milieu-management*, Samson/H. D. Tjeenk Willink, Alphen aan den Rijn.

Zoeteman, K. (1988), 'De aarde is ziek, de mens bedrijfsblind', *Intermediair*, 24/15: 69–75.

—— (1989), *Gaiasofie*, Ankh-Hermes, Deventer.

SECONDARY SOURCES

Ackermann, B. and Hassler, W. (1981), *Clean Coal, Dirty Air: Or how the Clean Air Act Became a Multibillion Dollar Bail-out for High-Sulphur Coal Producers and What Should be Done about it*, Yale UP, New Haven, Conn.

Amy, D. J. (1987), *The Politics of Environmental Mediation*, Columbia University Press, New York.

Anon. (1972), 'Britain Breathes Sweeter', *Nature*, 238 (18 Aug.), 366.

Anon. (1983b), 'Royal Society Appointed Referee in UK Dispute', *Nature*, 305 (8 Sept.), 85.

Anon. (1984), 'How to Neutralize Acid Rain', *Nature*, 310 (23 Aug.), 611–12.

Anon. (1986b), 'Voorlichting over verzuring door VROM', *Lucht en Omgeving* (Nov./Dec.), 154–6.

Ashby, E. and Anderson, M. (1976), 'Studies in the Politics of Environmental Protection: The Historical Roots of the British Clean Air Act, 1956: I. the Awakening of Public Opinion over Industrial Smoke, 1843–1853', *Interdisciplinary Science Reviews*, 1/4: 279–90.

—— —— (1981), *The Politics of Clean Air*, Clarendon Press, Oxford.

Ashby, K. R. (1982), 'Acid Rain', *Science and Public Policy* (Apr.), 106–10.

Austin, J. L. (1962), *How to do Things with Words*, 1992 edn., Oxford, University Press, Oxford.

Barmentlo, I. (1988), 'Shell als succesvol lobbyist', unpub. diss., University of Leiden.

Beck, U. (1986), *Risikogesellschaft—Auf dem Weg in eine andere Moderne*, Suhrkamp, Frankfurt am Main.

—— (1988), *Gegengifte—Die organisierte Unverantwortlichkeit*, Suhrkamp, Frankfurt am Main.

—— (1991), *Politik in der Risikogesellschaft*, Suhrkamp, Frankfurt am Main.

—— (1992), 'From Industrial Society to the Risk Society: Questions of Survival, Social Structure and Ecological Enlightenment', *Theory, Culture and Society*, 9: 97–123.

—— (1994), 'Self-Dissolution and Self-Endangerment of Industrial Society: What Does This Mean?', in Beck *et al.*, 174–83.

—— Lash, S., and Giddens, A. (1994), *Theories of Reflexive Modernisation*, Polity Press, Cambridge.

Beetham, D. (1991), *The Legitimation of Power*, Macmillan, London.

Benton, T. and Redclift, M. (1994), 'Introduction', in M. Redclift and T. Benton (eds.), *Social Theory and the Global Environment*, Routledge, London, 1–27.

Berg, J. T. J. van den, and Molleman, H. A. A. (1975), *Crisis in de Nederlandse politiek*, Samson, Alphen aan den Rijn.

Berger, P. and Luckmann, T. (1966), *The Social Construction of Reality— A Treatise in the Sociology of Knowledge*, 1984 edn., Penguin, Harmondsworth.

Bernstein, R. J. (1976), *The Restructuring of Social and Political Theory*, 1979 edn., Methuen, London.

Billig, M. (1987), *Arguing and Thinking—A Rhetorical Approach to Social Psychology*, Cambridge University Press, Cambridge.

—— Condor, S., Edwards, D., Gane, M., Middleton, D., and Radley, A. (1988), *Ideological Dilemmas: A Social Psychology of Everyday Thinking*, Sage, London.

Boehmer-Christiansen, S. (1988*a*), 'Black Mist and the Acid Rain: Science as Fig Leaf of Policy', *Political Quarterly*, 59/2: 145–60.

—— and Skea, J. (1991), *Acid Politics: Environmental and Energy Policies in Britain and Germany*, Belhaven, London.

Bramwell, A. (1989), *Ecology in the 20th Century: A History*, Yale University Press, New Haven, Conn.

Breemen, N. van, *et al.* (1982), 'Soil Acidification from Atmospheric Ammonium Sulphate in Forest Canopy Throughfall', *Nature*, 299 (7 Oct.), 548–50.

Brooks, D. B. (1992), 'The Challenge of Sustainability: Is Integrating Environment and Economics Enough?', *Policy Sciences*, 26: 401–8.

Burchell, G. *et al.* (1991), *The Foucault Effect—Studies in Governmentality*, Harvester Wheatsheaf, London.

Burns, T. R. and Ueberhorst, R. (1988), *Creative Democracy: Systematic Conflict Resolution and Policy-Making in a World of High Science and High Technology*, Praeger, New York.

Burstein, P. (1991), 'Policy Domains: Organization, Culture, and Policy Outcomes', *Annual Review of Sociology*, 17: 327–50.

Buttel, F. H. and Taylor, P. J. (1992), 'Environmental Sociology and Global Environmental Change: A Critical Assessment', *Society and Natural Resources*, 5: 211–30.

Caljé, K. (1989), 'De grenzen aan de groei zijn plotseling terug', *NRC Handelsblad*, 18 Jan. 1989, p. 5.

Callon, M. and Latour, B. (1981), 'Unscrewing the Big Leviathan: How Actors Macro-structure Reality and how Sociologists help Them to do so', in K. Knorr-Cetina and A. V. Cicourel (eds.), *Advances in Social Theory and Methodology: Toward an Integration of Micro and Macro Sociologies*, Routledge and Kegan Paul, Boston: 277–303.

Carson, R. (1962), *Silent Spring*, 1963 edn., Hamish Hamilton, London.

Chatterjee, P. and Finger, M. (1992), 'UNCED Follow Up: The Same Old Order', *Eco-Currents*, 2/3.

Cherfas, J. (1990), 'Greenpeace and Science: Oil and Water?', in *Science*, 247: 1288–90.

Clark, W. C. and Majone, G. (1985), 'The Critical Appraisal of Scientific Inquiries with Policy Implications', *Science, Technology and Human Values*, 10/3: 6–19.

—— and Munn, R. E. (1986), *Sustainable Development of the Biosphere*, Cambridge University Press/IIASA, Cambridge.

Cohen, J. L. (1985), 'Strategy or Identity: New Theoretical Paradigms and Contemporary Social Movements', *Social Research*, 52/4: 663–716.

Cohen, M. *et al.* (1972), 'A Garbage Can Model of Organisational Choice', *Administrative Science Quarterly*, 17: 1–25.

Cole, H. S. D., Freeman, C., Jahoda, M. and Pavitt, K. L. R. (1974), *Thinking about the Future: A Critique of The Limits to Growth*, Chatto & Windus/Sussex University Press, London.

Cotgrove, S. (1982), *Catastrophe or Cornucopia: The Environment, Politics and the Future*, Wiley, Chichester.

—— and Duff, A. (1980), 'Environmentalism, Middle-class Radicalism and Politics', *Sociological Review*, 28/2: 333–51.

Court, T. de la (1990), *Beyond Brundtland—Green Development in the 1990s*, Zed Books, London.

Cowling, E. B. (1982), 'Acid Rain in Historical Perspective', *Environmental Science and Technology*, 16/2: 110–23.

Cramer, J. (1990), 'New Environmentalism in the Netherlands', in A. Jamison *et al.* (eds.), *The Making of the New Environmental Consciousness*, Edinburgh University Press, Edinburgh, 121–84.

Crick, B. (1991), 'The English and the British', in B. Crick (ed.), *National*

Identities—The Constitution of the United Kingdom, Basil Blackwell, Oxford, 90–104.

Crouch, C. (1983), 'Market Failure: Fred Hirsch and the Case for Social Democracy', in A. Ellis and K. Kumar (eds.), 185–203.

Cuperus, R. (1992), 'Culturele Milieuschade?', *Socialisme and Democratie*, 49/10: 451–4.

Daalder, H. (1966), 'The Netherlands: Opposition in a Segmented Society', in R. A. Dahl (ed.), *Political Oppositions in Western Democracies*, Yale University Press, New Haven, Conn., 196–213.

Dahl, R. A. (1961), *Who Governs?*, Yale University Press, New Haven, Conn.

Dalton, R. J. (1988), *Citizen Politics in Western Democracies: Public Opinion in the United States, Great Britain, West Germany and France*, Chatham House, Chatham, NJ.

Daly, H. (1977), *Steady State Economics*, Freeman, San Francisco.

Davies, B. and Harré, R. (1990), 'Positioning: The Discursive Production of Selves', *Journal for the Theory of Social Behaviour*, 20/1: 43–63.

Dieleman, J. P. C. (1987), *Denken over Milieubeleid*, Erasmus Universiteit, Rotterdam.

Dijk, T. A. van (1985) (ed.), *Handbook of Discourse Analysis*, i. *Disciplines of Discourse*, Academic Press, London.

Distelbrink, F. *et al.* (1985), *Zure regen—Zelf schade herkennen aan bomen*, WISE, Amsterdam.

Dobson, A. (1990), *Green Political Thought*, Unwin Hyman, London.

Douglas, M. (1966), *Purity and Danger—An Analysis of the Concepts of Pollution and Taboo*, 1988 edn., Routledge, London.

—— (1982) (ed.), *Essays in the Sociology of Perception*, Routledge and Kegan Paul, London.

—— (1987), *How Institutions Think*, Routledge and Kegan Paul, London.

—— (1992), *Risk and Blame—Essays in Cultural Theory*, Routledge, London.

Downs, A. (1972), 'Up and Down with Ecology—The "Issue-Attention Cycle"', *Public Interest*, 28 (summer), 38–50.

Dreyfus, H. L. and Rabinow, P. (1986), *Michel Foucault—Beyond Structuralism and Hermeneutics*, Harvester, Brighton.

Dryzek, J. S. (1990a), *Discursive Democracy—Politics, Policy, and Political Science*, Cambridge University Press, Cambridge.

—— (1990b), 'Designs for Environmental Discourse: The Greening of the Administrative State?', in Paehlke and Torgerson (eds.): 97–111.

Dudley, N. *et al.* (1985), *The Acid Rain Controversy*, Earth Resources, London.

Dunlap, R. E., Kraft, M., and Rosa, E. A. (1993) (eds.), *Public Reactions to Nuclear Waste—Citizens' Views of Repository Siting*, Duke University Press, Durham, NC.

Dunn, W. N. (1993), 'Policy Reforms as Arguments', in F. Fischer and J. Forester (eds.), *The Argumentative Turn in Policy and Planning*, Duke University Press, Durham, NC., 254–90.

Eckersley, R. (1989), 'Green Politics and the New Class: Selfishness or Virtue?', in *Political Studies*, 17: 205–23.

—— (1992), *Environmentalism and Political Theory: Towards an Ecocentric Approach*, UCL Press, London.

Edelman, M. (1971), *Politics as Symbolic Action—Mass Arousal and Quiescence*, Academic Press, New York.

Edwards, D. and Potter, J. (1992), *Discursive Psychology*, Sage, London.

Ekins, P. (1992), *A New World Order—Grassroots Movements for Global Change*, Routledge, London.

Ellis, A. and Kumar, K. (1983) (eds.), *Dilemmas of Liberal Democracies: Studies in Fred Hirsch's* Social Limits to Growth, Tavistock, London.

Elsworth, S. (1984), *Acid Rain*, Pluto Press, London.

Engbersen, R. (1991), 'Profane Boeken der Openbaring', in Engbersen *et al.* (eds.): 145–64.

Engbersen, R. *et al.* (1991) (eds.), *Het Retorische Antwoord*, Grafiet, Utrecht.

Environmental Resources Limited (1983), *Acid Rain—A Review of the Phenomenon in the EEC and Europe*, Graham & Trotman, London.

Enzensberger, H. M. (1973), 'Zur Kritik der politischen Oekologie', *Kursbuch*, 33: 1–42.

Everden, N. (1992), *The Social Creation of Nature*, Johns Hopkins University Press, Baltimore.

Ewald, F. (1986), *L'Etat Providence*, Grasset, Paris.

Finger, M. (1993), 'Politics of the UNCED Process', in W. Sachs (ed.), *Global Ecology: A New Arena of Political Conflict*, Zed Books, London: 36–48. *("radical" ?)*

Fischer, F. (1980), *Politics, Values, and Public Policy: The Problem of Methodology*, Westview, Boulder, Colo.

—— (1990a), *Technocracy and the Politics of Expertise*, Sage, London.

—— (1990b), *Rethinking Risk Assessment—Toward an Integration of Science and Participation*, Working Paper 32, Leiden Institute for Law and Public Policy, Leiden.

Fischhoff, B., Lichtenstein, S., Slovic, P., Derby, S. L., and Keeney, R. L. (1981), *Acceptable Risk*, Cambridge University Press, Cambridge.

Flam, H. (1994) (ed.), *States and Anti-Nuclear Movements*, Edinburgh University Press, Edinburgh.

Fleck, L. (1979), *Genesis and Development of a Scientific Fact*, University of Chicago Press, Chicago.

Forester, J. (1982), 'A Critical Empirical Framework for the Analysis of Public Policy', in *New Political Science*, 9/10: 33–61.

—— and Fischer, F. (1993) (eds.), *The Argumentative Turn in Policy Analysis and Planning*, Duke University Press, Durham, NC.

Forrester, J. W. (1969), *Urban Dynamics*, MIT Press, Cambridge, Mass.
—— (1971), *World Dynamics*, Wright-Allen, Cambridge, Mass.
Foucault, M. (1966), *The Order of Things—An Archaeology of the Human Sciences*, 1973 trans., Vintage Books, New York.
—— (1968), 'Politics and the Study of Discourse', repr. in Burchell *et al.* (1991), 53–72.
—— (1971), *L'ordre du discours*, Gallimard, Paris.
—— (1976), *The History of Sexuality*, i, 1981 trans., Penguin, Harmondsworth.
—— (1975), *Discipline and Punish—The Birth of the Prison*, 1991 trans., Penguin, Harmondsworth.
—— (1980), 'Powers and Strategies', in C. Gordon (ed.), *Power/Knowledge*, Pantheon Books, New York, 134–45.
Galbraith, J. K. (1967), *The New Industrial State*, 1979 edn., Penguin, Harmondsworth.
Galtung, J. (1986), 'The Green Movement: A Socio-Historical Exploration', *International Sociology*, 1/1: 75–90.
Gamble, D. J. (1987), 'The Berger Inquiry', *Science*, 199: 946–52.
Gamson, W. A. and Modigliani, A. (1989), 'Media Discourse and Public Opinion on Nuclear Power: A Constructionist Approach', *American Journal of Sociology*, 95/1: 1–37.
Georgescu-Roegen, N. (1974), *The Entropy Law and Economic Progress*, Harvard University Press, Cambridge, Mass.
Gershuny, J. (1983), 'Technical Change and "Social Limits"', in Ellis and Kumar (1983), 23–44.
Gibbons, M. T. (1987) (ed.), *Interpreting Politics*, Blackwell, Oxford.
Gibson, J. (1991), 'The Integration of Pollution Control', *Journal of Law and Society*, 18/1: 18–31.
Giddens, A. (1979), *Central Problems in Social Theory—Action, Structure and Contradiction in Social Analysis*, University of California Press, Berkeley.
—— (1984), *The Constitution of Society*, Polity Press, Cambridge.
—— (1990), *The Consequences of Modernity*, Polity Press, Cambridge.
Glacken, C. J. (1967), *Traces on the Rhodian Shore*, University of California Press, Berkeley.
Golub, R. and Townsend, J. (1977), 'Malthus, Multinationals and the Club of Rome', in *Social Studies of Science*, 7: 201–22.
Goodin, R. A., *Green Political Theory*, Polity Press, Cambridge.
Gordon, C. (1991), 'Governmental Rationality: An Introduction', in Burchell *et al.* (1991), 1–51.
Gouldner, A. W. (1979), *The Future of Intellectuals and the Rise of the New Class*, Macmillan, London.
Gravensteyn, L. J. J. (1984), Het IMP-Lucht 1985–1989, *Lucht en Omgeving*, Sept./Oct., 144–8.

Gray, J. S. (1990), 'Statistics and the Precautionary Principle', *Marine Pollution Bulletin*, 21/4: 174–6.

Gusfield, J. R. (1981), *The Culture of Public Problems: Drinking-Driving and the Symbolic Order*, University of Chicago Press, Chicago.

Haas, P. M. (1990), *Saving the Mediterranean—The Politics of International Environmental Cooperation*, Columbia University Press, New York.

Habermas, J. (1981), *Theorie des kommunikativen Handelns*, i, ii, Suhrkamp, Frankfurt.

—— (1986), *Der Philosophische Diskurs der Moderne*, Suhrkamp, Frankfurt am Main.

Haigh, N. (1987), *EEC Environmental Policy and Britain*, Longman, Harlow.

Hajer, M. A. (1990), 'The Discursive Paradox of the New Environmentalism', *Industrial Crisis Quarterly*, 4/4: 307–10.

—— (1992*a*), 'Milieu: een kwestie van risico's?', in N. J. H. Huls (ed.), *Sturing in de risicomaatschappij*, W. E. J. Tjeenk Willink, Zwolle, 115–36.

—— (1992*b*), 'Milieu als moderniseringsvraagstuk', *Socialisme & Democratie*, 49/1: 1–7.

—— (forthcoming), 'Ecological Modernisation and Social Change', in S. Lash, B. Szerzinski, and B. Wynne (eds.), *Risk, Environment and Modernity: Towards a New Ecology*, Sage, London.

Hanf, K. (1989), 'Deregulation as Regulatory Reform: The Case of Environmental Policy in the Netherlands', *European Journal of Political Research*, 17: 193–207.

Harré, R. (1993), *Social Being*, Blackwell, Oxford.

Harvey, D. (1974), 'Population, Resources and the Ideology of Science', in *Economic Geography*, 50: 256–77.

—— (1989), *The Condition of Postmodernity: An Enquiry into the Origins of Cultural Change*, Blackwell, Oxford.

—— (1993), 'The Nature of Environment: The Dialectics of Social and Environmental Change', in R. Miliband and L. Panitch (eds.), *The Socialist Register*, Merlin Press, London: 1–51.

Hawkins, K. (1984), *Environment and Enforcement—Regulation and the Social Definition of Pollution*, Clarendon Press, Oxford.

Hays, S. (1959), *Conservation and the Gospel of Efficiency—The Progressive Conservation Movement 1890–1920*, 1979 edn., Atheneum, New York.

Heijden, H. A. van der (1992), 'Van kleinschalig utopisme naar postgiroactivisme? De milieubeweging 1970–1990', in J. W. Duyvendak *et al.*, *Tussen verbeelding en macht, 25 jaar sociale bewegingen in Nederland*, SUA, Amsterdam, 77–98.

Hemerijck, A. C. (1993), 'The Historical Contingencies of Dutch Corporatism', D.Phil thesis, University of Oxford.

Hildebrandt, E., Gerhardt, U., Kühleis, C., Schenk, S., and Zimpelmann, B. (1994), 'Politisierung und Entgrenzung—am Beispiel ökologisch erweiterter Arbeitspolitik', in N. Beckenbach and N. van Treeck (eds.), *Umbrüche gesellschaftlicher Arbeit*, Otto Schwartz & Co., Göttingen: 429–44.

Hirsch, F. (1977), *Social Limits to Growth*, Routledge & Kegan Paul, London.

Holdgate, M. W. (1979), *A Perspective of Environmental Pollution*, Cambridge University Press, Cambridge.

Honigh, M. (1985), *Doeltreffend beleid: Een empirische vergelijking van beleidssectoren*, Van Gorcum, Assen.

Hoppe, R. (1983), *Economische zaken schrijft een nota—een onderzoek naar beleidsontwikkeling en besluitvorming by nonincrementeel beleid*, VU uitgeverij, Amsterdam.

Huber, J. (1982), *Die verlorene Unschuld der Oekologie*, S. Fischer, Frankfurt am Main.

——— (1985), *Die Regenbogengesellschaft. Oekologie und Sozialpolitik*, S. Fischer, Frankfurt am Main.

——— (1991), *Unternehmen Umwelt. Weichenstellungen für eine ökologische Marktwirtschaft*, S. Fischer, Frankfurt am Main.

Huisingh, D. and Bailey, V. (1982) (eds.), *Making Pollution Prevention Pay—Ecology with Economy as Policy*, Elmsford, New York.

——— et al. (1986), *Proven Profits from Pollution Prevention—Case Studies in Resource Conservation and Waste Reduction*, Washington, DC.

Inglehart, R. (1971), 'The Silent Revolution in Europe: Intergenerational Change in Post-Industrial Societies', in *American Political Science Review*, 65: 991–1017.

——— (1977), *The Silent Revolution: Changing Values and Political Styles among Western Publics*, Princeton University Press, Princeton, NJ.

——— (1984), 'Traditionelle politische Trennungslinien und die Entwicklung der neuen Politik in westlichen Gesellschaften', in *Politische Vierteljahresschriften*, 2: 139–65.

——— (1990), *Culture Shift in Advanced Democracies*, Princeton University Press, Princeton, NJ.

Jänicke, M. (1985), *Preventive Environmental Policy as Ecological Modernisation and Structural Policy*, WZB, Berlin.

——— (1988), 'Oekologische Modernisierung—Optionen und Restriktionen präventiver Umweltpolitik', in U. E. Simonis (ed.), 13–26.

Jasanoff, S. (1986), *Risk Management and Political Culture*, Russell Sage Foundation, New York.

——— (1990), *The Fifth Branch: Science Advisors as Policymakers*, Harvard University Press, Cambridge, Mass.

Johnston, P. et al. (1988), 'Greenpeace and the East Enders', *Scope* (winter), 27–30.

Jordan, A. G. and Richardson, J. J. (1987), *British Politics and the Policy Process*, Allen & Unwin, London.

Jungk, R. (1977), *Der Atomstaat: Vom Fortschritt in die Unmenschlichkeit*, Rororo, München.

Kapp, K. W. (1950), *The Social Costs of Private Enterprise*, Harvard University Press, Cambridge, Mass.

Katzenstein, P. (1985), *Small States in World Markets: Industrial Policy in Europe*, Cornell University Press, Ithaca, NY.

Keller, R. and Poferl, A. (1994), 'Habermas und Müll', *Wechselwirkung*, 16/68: 34–40.

Kingdon, J. W. (1984), *Agendas, Alternatives, and Public Policies*, Harper Collins, Glasgow.

Kneese, A., Ayres, R., and d'Arge, R. (1971), *Economics and the Environment: A Materials Balance Approach*, Johns Hopkins University Press, Baltimore.

Knoepfel, P. and Weidner, H. (1986), 'Explaining Differences in the Performances of Clean Air Policies: An International and Interregional Comparative Study', *Policy and Politics*, 14/1: 71–91.

Kuhn, T. S. (1962), *The Structure of Scientific Revolutions*, 1970 edn., University of Chicago Press, Chicago.

Kwa, C. (1987), 'Representations of Nature mediating between Ecology and Science Policy: The Case of the International Biological Programme', *Social Studies of Science*, 17: 413–42.

—— (1994), 'De Symbolische Beheersing van de Wereld', *ZENO*, 2/3: 26–7.

Lakoff, G. and Johnson, M. (1980), *Metaphors We Live By*, University of Chicago Press, Chicago.

Leemans, A. F. and Geers, K. (1983), *Doorbraak in het Oosterschelde-beleid*, Coutinho, Muiderberg.

Leiss, W. (1978), *The Limits to Satisfaction—On Needs and Commodities*, Marion Boyars, London.

Lembruch, G. (1984), 'Concertation and the Structure of Corporatist Networks', in J. H. Goldthorpe (ed.), *Order and Conflict in Contemporary Capitalism*, Clarendon Press, Oxford, 60–80.

Leroy, P. (1985), *Milieubeweging en milieubeleid*, Nederlandse Boekhandel, Antwerpen.

Liberatore, A. (1994), 'Facing Global Warming—The Interactions between Science and Policy-making in the European Community', in Redclift and Benton (eds.), 190–204.

Liefferink, J. D., Lowe, P., and Mol, A. P. J. (1993) (eds.), *European Integration and Environmental Policy*, Belhaven, London.

Lindblom, C. E. (1977), *Politics and Markets—The World's Political Economic Systems*, Basic Books, New York.

Lloyd, C. (1986), *Explanation in Social History*, Blackwell, Oxford.

Loo, H. van der (1992), 'De zegetocht van het publieke debat', *De Helling*, 5/2: 23–7.

Lowe, P. and Goyder, J. (1983), *Environmental Groups in Politics*, Allen & Unwin, London.

—— and Rüdig, W. (1986), 'Review Article: Political Ecology and the Social Sciences—The State of the Art', in *British Journal of Political Science*, 16: 513–50.

Luhmann, N. (1984), *Soziale Systeme: Grundriss einer allgemeinen Theorie*, Suhrkamp, Frankfurt am Main.

—— (1986), *Oekologische Kommunikation—Kann die moderne Gesellschaft sich auf ökologische Gefahrdungen einstellen?*, Westdeutscher Verlag, Opladen.

Lukes, S. (1974), *Power—A Radical View*, Methuen, London.

Lundgren, L. J. (1991), *Försurningen pa Dagordningen*, Naturvardsverket, Solna.

Lykke, E. (1992) (ed.), *Achieving Environmental Goals—The Concept and Practice of Environmental Performance Review*, Belhaven, London.

Lijphart, A. (1968), *The Politics of Accommodation: Pluralism and Democracy in the Netherlands*, University of California Press, Berkeley.

McCormick, J. (1990), *Acid Earth: The Global Threat of Acid Pollution* (rev. edn.), Earthscan/WWF, London.

McLellan, D. (1986), *Ideology*, Open University Press, Milton Keynes.

MacLeod, R. M. (1965), 'The Alkali Acts Administration 1863–84: The Emergence of the Civil Scientist', *Victorian Studies*, 912: 85–112.

Macrory, R. (1986), *Environmental Policy in Britain: Reaffirmation or Reform?*, WZB-paper IIUG 86–4, WZB, Berlin.

Majone, G. (1989a), *Evidence, Argument and Persuasion in the Policy Process*, Yale University Press, New Haven, Conn.

—— (1989b) (ed.), *Deregulation or Regulation? Regulatory Reform in Europe and the United States*, Pinter, London.

Mannheim, K. (1940), *Man and Society in an Age of Reconstruction—Studies in Modern Social Structure*, 1960 edn., Routledge & Kegan Paul, London.

Marquand, D. (1988), *The Unprincipled Society—New Demands and Old Politics*, Fontana, London.

Marsh, G. P. (1864), *Man and Nature; or, Physical Geography as Modified by Human Action*, Scribner, New York.

Maslow, A. H. (1954), *Motivation and Personality*, Harper & Row, New York.

Masters, D. and Way, K. (1946) (eds.), *One World or None: A Report to the Public on the Full Meaning of the Atomic Bomb*, McGraw Hill, New York.

Melucci, A. (1985), 'The Symbolic Challenge of Contemporary Movements', *Social Research*, 789–816.

Merton, R. K. (1970), *Science, Technology and Society in Seventeenth Century England*, Harvester Press, Sussex.

—— (1973), *The Sociology of Science—Theoretical and Empirical Investigations*, Chicago University Press, Chicago.

Middendorp, C. P. (1979), *Ontzuiling, politisering en restauratie in Nederland—de jaren 60 en 70*, Boom, Meppel.

Nicholson, I. A. *et al.* (1980), 'pH and Sulphate Content of Precipitation over Northern Britain', *Ecological Impact of Acid Precipitation*, SNSF, Oslo, 142–3.

Offe, C. (1985), 'New Social Movements: Challenging the Boundaries of Institutional Politics', *Social Research*, 52/4 (winter), 817–68.

O'Riordan, T. (1983), *Environmentalism*, Pion Press, London.

—— (1985), 'Culture and the Environment in Britain', *Environmental Management*, 9/2: 113–20.

—— (1988), 'Anticipatory Environmental Policy: Impediments and Opportunities', in Simonis (ed.), 65–76.

—— and Weale, A. (1989), 'Administrative Reorganization and Policy Change: The Case of Her Majesty's Inspectorate of Pollution', *Public Administration*, 67: 277–94.

Paehlke, R. C. (1989), *Environmentalism and the Future of Progressive Politics*, Yale University Press, New Haven, Conn.

—— and Torgerson, D. (1990) (eds.), *Managing Leviathan—Environmental Politics and the Administrative State*, Broadview, Peterborough (Ont.).

Park, C. C. (1986), *Acid Rain—Rhetoric and Reality*, Routledge, London.

Parker, R. (1975), 'The Struggle for Clean Air', in P. Hall *et al.*, *Change, Choice and Conflict in Social Policy*, Heinemann, London, 371–409.

Pearce, F. (1982), 'It's an Acid Wind that Blows Nobody any Good', *New Scientist* (8 July), 80.

—— (1986), 'Unravelling a Century of Acid Pollution', *New Scientist* (25 Sept.), 23–4.

—— (1987), *Acid Rain*, Penguin, Harmondsworth.

Pearce, D., MarKandya, A., and Barbier, E. B. (1989), *Blueprint for a Green Economy*, Earthscan, London.

Pepper, D. (1986), *The Roots of Modern Environmentalism*, Routledge, London.

Perrow, C. (1984), *Normal Accidents: Living with High-Risk Technologies*, Basic Books, New York.

Peters, T. J. and Waterman, R. H. (1982), *In Search of Excellence—Lessons from America's Best Run Companies*, Harpers & Row, New York.

Polsby, N. W. (1963), *Community Power and Political Theory*, Yale University Press, New Haven, Conn.

Porritt, J. and Winner, D. (1988), *The Coming of the Greens*, Fontana, London.

Potter, J. and Wetherell, M. (1987), *Discourse and Social Psychology—Beyond Attitudes and Behaviour*, Sage, London.

Prittwitz, V. von (1983), 'Europäische Zusammenarbeit in der Luftreinhaltung?', *Zeitschrift für Umweltpolitik*, 2: 117–32.

—— (1990), *Das Katastrophenparadox—Elementen einer Theorie der Umweltpolitik*, Leske & Budrich, Opladen.

Pullen, M. P. J. and Wiersma, D. (1986), 'Bestrijdingsmogelijkheden SO_2-emissie door raffinaderijen', *Lucht & Omgeving*, Jan./Feb., 14–17.

Rand Hoare, M. (1991), 'When the Earth Moved', *Times Higher Education Supplement*, 1 Jan., 15.

Ravetz, J. R. (1971), *Scientific Knowledge and its Social Problems*, Oxford University Press, Oxford.

—— (1990), *The Merger of Knowledge with Power—Essays in Critical Science*, Mansell, London.

Redclift, M. (1987), *Sustainable Development—Exploring the Contradictions*, Methuen, London.

Reed, L. E. (1976), 'The Long-range Transport of Air Pollutants', *Ambio*, 5: 164–8.

Regens, J. L. and Rycroft, R. W. (1988), *The Acid Rain Controversy*, University of Pittsburgh Press, Pittsburgh.

Rein, M. and Schön, D. A. (1986), 'Frame-Reflective Policy Discourse', *Beleidsanalyse*, 15/4: 4–18.

Richardson, J. J. (ed.) (1982), *Policy Styles in Western Europe*, George Allen & Unwin, London.

—— Jordan, A. G. (1979), *Governing under Pressure—The Policy Process in a Post-Parliamentary Democracy*, Martin Robertson, Oxford.

Rittel, H. W. J. and Webber, M. M. (1973), 'Dilemmas in a General Theory of Planning', *Policy Sciences*, 4: 155–69.

Rohrschneider, R. (1988), 'Citizens' Attitudes toward Environmental Issues: Selfish or Selfless?', in *Comparative Political Studies*, 21/3: 347–67.

Roqueplo, P. (1986), 'Der Saure Regen: Ein "Unfall in Zeitluppe"—Ein Beitrag zu einer Soziologie des Risikos', *Soziale Welt*, 4: 402–26.

Roszak, T. (1969), *The Making of a Counter Culture—Reflections on the Technocratic Society and its Youthful Opposition*, Anchor Books, New York.

Royston, M. G. (1979), *Pollution Prevention Pays*, Pergamon Press, Oxford.

Rüdig, W. (1988), 'Peace and Ecology Movements in Western Europe', in *West European Politics*, 11/1: 26–39.

—— (1990), *Anti-Nuclear Movements—A World Survey of Opposition to Nuclear Energy*, Longman, Harlow.

—— (1994), 'Maintaining a Low Profile: The Anti-Nuclear Movement and the British State', in Flam (ed.), 70–100.

Sabatier, P. A. (1987), 'Knowledge, Policy-oriented Learning, and Policy Change', *Knowledge: Creation, Diffusion, Utilization*, 8/4: 649–62.

Sachs, W. (1992), 'Environment', in W. Sachs (ed.), 26–37.

—— (1992) (ed.), *The Development Dictionary—A Guide to Knowledge as Power*, Zed Books, London.

Sandbach, F. (1978), 'The Rise and Fall of the *Limits to Growth* Debate', in *Social Studies of Science*, 8: 495–520.

Sanderson, J. B. (1961), 'The National Smoke Abatement Society and the Clean Air Act (1956)', *Political Studies*, 9/3: 236–53.

Schama, S. (1991), *The Embarrassment of Riches—An Interpretation of Dutch Culture in the Golden Age*, Fontana, London.

Schattschneider, E. E. (1960), *The Semi-Sovereign People*, Rinehart and Winston, New York.

Schmandt, J., Clarkson, J., and Roderick, H. (1988) (eds.), *Acid Rain and Friendly Neighbors—The Policy Dispute between Canada and the United States*, Duke University Press, Durham, NC.

Schön, D. A. (1979), 'Generative Metaphors in the Setting of Social Policy Problems', in A. Ortony (ed.), *Metaphor and Thought*, Cambridge University Press, Cambridge.

Schumpeter, J. A. (1943), *Capitalism, Socialism and Democracy*, 1961 edn., George Allen and Unwin, London.

Schütt, P., Koch, W., Schück, H. J., Lang, K. J., and Summerer, H. (1983), *So Stirbt der Wald*, BLV, München.

Schwarz, M. and Thompson, M. (1990), *Divided We Stand—Redefining Politics, Technology and Social Choice*, Harvester Wheatsheaf, London.

Scimemi, G. (1988), 'Environmental Policies and Anticipatory Strategies', in Simonis (ed.), 27–48.

Shackley, S., Wynne, B., Parkinson, S., Young, P. (1993), *Mission to Model Earth*, CSEC/CRES Joint Paper, University of Lancaster.

Shapin, S. and Schaffer, S. (1985), *Leviathan and the Air-pump—Hobbes, Boyle, and the Experimental Life*, Princeton University Press, Princeton, NJ.

Shiva, V. (1993), 'The Greening of the Global Reach', in Sachs (ed.): 149–56.

Simmons, I. G. (1989), *Changing the Face of the Earth—Culture, Environment, History*, Blackwell, Oxford.

Simonis, U. E. (1988) (ed.), *Präventive Umweltpolitik*, Campus Verlag, Frankfurt am Main.

Spaargaren, G. and Mol, A. P. J. (1992), 'Sociology, Environment, and Modernity: Ecological Modernization as a Theory of Social Change', *Society and Natural Resources*, 5: 323–44.

Straaten, J. van der (1990), *Zure Regen, Economische Theorie en het Nederlandse Beleid*, Jan van Arkel, Utrecht.

Stuurman, S. (1983), *Verzuiling, kapitalisme en patriarchaat*, SUN, Nijmegen.

Taylor, C. (1972), 'The Politics of the Steady State', in A. Rotstein (ed.), *Beyond Industrial Growth*, University of Toronto Press, Toronto, 47–70.

Tellegen, E. (1979), 'Oude en Nieuwe Milieuorganisaties', in P. Ester, *Sociale Aspecten van het Milieuvraagstuk*, Van Gorcum, Assen, 152–68.

—— (1989), 'De onmacht in kaart gebracht: over het nationaal milieubeleidsplan', *Rooilijn*, 7: 200–4.

Teubner, G. (1992), *The Invisible Cupola—From Causal to Collective Attribution in Ecological Liability*, European University Institute Working Paper 51/92.

—— Farmer, L., and Murphy, D. (1994) (eds.), *Environmental Law and Ecological Responsibility: The Concept and Practice of Ecological Self-Organization*, John Wiley, London.

Thomas, K. (1983), *Man and the Natural World—Changing Attitudes in England 1500–1800*, Penguin, Harmondsworth.

Thomas, W. L. (1956) (ed.), *Man's Role in Changing the Face of the Earth*, University of Chicago Press, Chicago.

Thompson, M., Ellis, R., and Wildavsky, A. (1990), *Cutural Theory*, Westview Press, Boulder, Colo.

Tinbergen, J. (1977), *RIO: Reshaping the International Order—A Report to the Club of Rome*, New American Library, New York.

Toffler, A. (1980), *The Third Wave*, Morrow, New York.

Torgerson, D. (1990), 'Limits of the Administrative Mind: The Problem of Defining Environmental Problems', in Paehlke and Torgerson (eds.), 115–61.

Turner, B. L. (1991) (ed.), *The Earth as Transformed by Human Action: Global and Regional Changes in the Biosphere over the past 300 Years*, Cambridge University Press, Cambridge.

Vogel, D. (1986), *National Styles of Regulation: Environmental Policy in Great Britain and the United States*. Cornell University Press, Ithaca, NY.

Wagner, P. (1990), *Sozialwissenschaften und Staat—Frankreich, Italien, Deutschland 1870–1980*, Campus, Frankfurt.

—— and Wittrock, B. (1987), *Social Sciences and Societal Developments: The Missing Perspective*, WZB paper P 87–4, WZB, Berlin.

Walgate, R. (1984a), 'Allies for UK Government', *Nature*, 310 (16 Aug.), 535.

—— (1984b), 'CEGB takes a pasting from MPs', *Nature*, 311 (13 Sept.), 94.

—— (1986), 'UK Denies Responsibility for Scandinavian Acid Rain', *Nature*, 323 (18 Sept.), 191.

Watts, N. and Wandesforde-Smith, G. (1981), 'Postmaterial Values and Environmental Policy Change', in D. E. Mann (ed.), *Environmental Policy Formation: The Impact of Values, Ideology, and Standards*, Lexington Books, Lexington, Mass., 29–42.

Weale, A. (1992), *The New Politics of Pollution*, Manchester University Press, Manchester.

Weber, M. 1978, *Economy and Society—An Outline of Interpretive Sociology*, i, University of California Press, Berkeley.

Weidner, H. (1986), '17 Länder im Vergleich—Luftreinhaltepolitik in Europa: Leistungen und Möglichkeiten', Pt. 1, *Umweltmagazin*, 15/7: 26–33.

—— (1987), *Clean Air Policy in Europe: A Survey of Regulations, Problems and Abatement Measures in 21 Countries*, Edition Sigma, Berlin.

—— and Knoepfel, P. (1985a), *Luftreinhaltepolitik bei stationären Quellen in England*, iii, Sigma, Berlin.

—— —— (1985b), *Luftreinhaltepolitik bei stationären Quellen in den Niederlanden*, vi, Sigma, Berlin.

Weinberg, A. M. (1972), 'Science and Trans-science', *Minerva*, 10/2: 209–22.

Wetherell, M. and Potter, J. (1988), 'Discourse Analysis and the Identification of Interpretive Repertoires', in C. Antaki (ed.), *Analysing Everyday Explanation*, Sage, London: 168–83.

Wetstone, G. S. (1987), 'A History of the Acid Rain Issue', in H. Brooks and C. L. Cooper, *Science for Public Policy*, Pergamon Press, Oxford, 163–95.

—— and Rozencranz, A. (1983), 'Transboundary Air Pollution in Europe: A Survey of National Responses', *Columbia Journal of Environmental Law*, 9/1: 1–62.

White, L. (1967), 'The Historical Roots of Our Ecological Crisis', *Science*, 155/3767 (10 Mar.) 1203–7.

Williams, R. (1973), *The Country and the City*, 1985 edn., Hogarth Press, London.

Witteveen, W. J. (1991), *Evenwicht van Machten*, W. E. J. Tjeenk Willink, Zwolle.

Wittrock, B., Wagner, P., and Wollmann, H. (1987), *Social Science and the Modern State: Knowledge, Institutions, and Societal Transformations*, WZB paper P 87–3, WZB, Berlin.

Worster, D. (1977), *Nature's Economy—A History of Ecological Ideas*, 1991 edn., Cambridge University Press, Cambridge.

Wynne, B. (1982), *Rationality and Ritual: The Windscale Inquiry and Nuclear Decisions in Britain*, The British Society for the History of Science, Chalfont St Giles.

—— (1987), *Risk Management and Hazardous Waste—Implementation and the Dialectics of Credibility*, Springer, Berlin.

Wynne, B. (1992*a*), 'Carving Out Science (and Politics) in the Regulatory Jungle' (Essay Review), *Social Studies of Science*, 22/3: 745–58.

—— (1992*b*), 'Misunderstood Misunderstandings: Social Identities and Public Uptakes of Science', *Public Understanding of Science*, 1/3: 281–304.

—— (1994), 'Scientific Knowledge and the Global Environment', in Redclift and Benton (eds.), 169–89.

Yearley, S. (1991), *The Green Case—A Sociology of Environmental Issues, Arguments and Politics*, Harper Collins, London.

Zahn, E. (1991), *Regenten, Rebellen en Reformatoren—Een visie op Nederland en de Nederlanders*, Contact, Amsterdam.

Zilleßen, H. (1993), 'Die Modernisierung der Demokratie im Zeichen der Umweltproblematik', in H. Zilleßen, P. C. Dienel, and W. Strubelt (eds.), *Die Modernisierung der Demokratie: Internationale Ansätze*, Westdeutscher Verlag, Opladen: 17–39.

Zizek, S. (1993), *Tarrying with the Negative: Kant, Hegel and the Critique of Ideology*, Duke University Press, Durham, NC.

INDEX